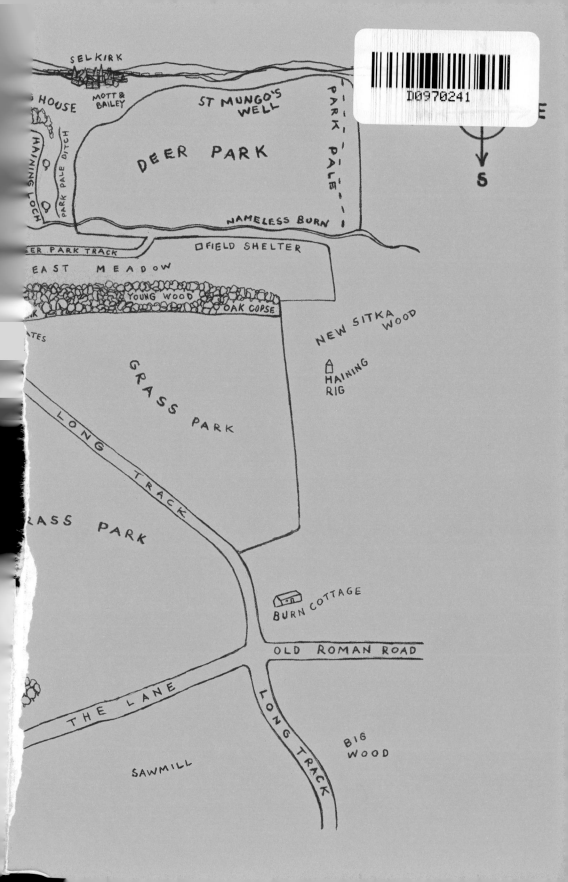

THE
SECRET
HISTORY
OF HERE

Also by Alistair Moffat

THE
SECRET
HISTORY
OF HERE

A YEAR IN THE VALLEY

ALISTAIR MOFFAT

CANONGATE

First published in Great Britain, the USA and Canada in 2021
by Canongate Books Ltd, 14 High Street, Edinburgh EH1 1TE

Distributed in the USA by Publishers Group West and in
Canada by Publishers Group Canada

canongate.co.uk

1

British Library Cataloguing-in-Publication Data
A catalogue record for this book is available on
request from the British Library

ISBN 978 1 83885 113 2

Typeset in Dante MT Std by
Palimpsest Book Production Ltd,
Falkirk, Stirlingshire

Printed and bound in Great Britain by Clays Ltd, Elcograf S.p.A.

For Walter Elliot and Rory Low

Foreword

Once Upon a Time

The Long Track lies fast asleep. The dawn track, it waits under the blanket of the night for the grey of the eastern sky to glow pink. Nothing stirs. Birds roost in the trees and the animals of the woods hide snug in their burrows, invisible to the blinking stare of the owls, perched still on the upper branches, watching, waiting for movement. The sleeping voles, rabbits and mice curl up in their own body warmth, safe from the snuffling of night predators nosing through the leaves of long-dead summers.

High above the Long Track the silence shudders with the wing-beats of two swans, their heads and necks spear-straight as they wheel south towards the waters of the great river. Under the shadows of the trees the night slumbers on.

Over the heads of the woods to the east the black horizon of the hills waits for the light to rise. In the half-world before dawn, the ghost hours, a sudden mist falls, shrouding the track and the trees. Far in the distance, roosting crows lift into the air, their cawing, cackling racket echoing around the valley. Then there is a snort, a whinny and out of the mist rides an armoured knight. Neck-reining his mighty destrier, he slows the stallion to a walk and soothing words murmur across the fields. Ramrod-straight in the saddle, his helm slung over the pommel, his armour a dull lustre in the morning light, the knight stares fixedly ahead. When a breeze suddenly riffles through the trees, the long caparison on his destrier billows, a blood-red crusader cross on white silk.

Behind rides an esquire on a little palfrey, his pennant planted in a stirrup niche, his liveried tabard bright with the device of his lord. Swivelling in the saddle, he turns to look for those who follow, listening for the creaks of the carts and the rumble of their wheels on the stones of the Long Track. The plod of the destrier and the palfrey linger as the knight and his esquire disappear into the morning mist.

On the edges of the wood, the children of the mist are playing. Visible only when they move, deer graze in the clearings. While the little fawns race each other, coming to sudden stops, making impossible turns, the does look up, ears pricked, searching for threat. In deep cover, it is waiting. Downwind, by the trickle of the burn at the foot of the track, two hunters hide in the tangle of undergrowth, silently grimacing at the prick of briars and thorns, camouflaged in their buckskin tunics. Moving very slowly, they inch their way closer to the grazing deer. They know they will have only one shot before their prey scatter, skittering back into the wood in a moment. The wind suddenly shifts, the does look up, their ears flicking, searching for the small sounds of danger, ready to run. The hunters nock their flint-tipped arrows, stand up, let fly, and a fawn staggers and stumbles. Pierced in the flank, the young animal bays in shock and pain, gets up and runs unsteadily. Even though the fawn gains the cover of the trees, the hunters will follow the blood trail and make their kill when the wounded animal is too weak to run beyond the range of their arrows.

Over the eastern ridges the sun climbs and the mist fades in moments. A tide of butter-coloured light washes down the valley, and on the Long Track the jingle of harness is heard. Wrapped in their cloaks against the morning chill, two riders trot up to the northern ridge. Without shields or spears, these scouts carry only the spatha, the cavalry sword. In hostile country, their eyes and ears are better weapons. Down in the valley, on the southern side of the burn, the ranks of a vast army march in battle order, their *nailed boots thudding, standards glinting red and gold in the morning sun. Trumpeters send signals: halt, stand easy, form line, close ranks, march on. Brigaded together into an invasion force, the legions of Gnaeus Julius Agricola, governor of the province of Britannia, have been sent into the territory of the Selgovae, a native kindred. Leading each detachment are the eagles of the XX Valeria Victrix, the IX Hispana and the II Augusta. They begin to climb into the western hills until their surveyors call a halt and mark out the perimeter of a camp. As a screen of sentries scan the hills for movement, ditches are quickly dug and at the gates a wooden tower raised. Far to the east, soldiers can see the flash of shiny metal from the signal station above the great army depot by the River Tweed. Wrapping the landscape in a web of roads, Rome tightens its grip. Hidden in the Wildwood, native warriors watch and wait.*

The shadow of war was not often cast over the Long Track. Few heard the clangour of distant battles. The march of history across the landscape usually happened elsewhere. Mostly the track was walked and ridden by peaceful travellers going about their business, or people seeking pleasure, perhaps solace. Bands of pilgrims bound for the shrines at Melrose or Jedburgh sang psalms, the burgesses of the royal burgh of Selkirk beat the bounds of their precious common land, wealthy estate owners planted an avenue of trees on either side of the track and made good the ruts and puddles so that their ladies could enjoy carriage drives, their umbrellas shading the afternoon sun. But war did change the track. In 1940, hardwood was needed and sawyers came to cut the mighty trees. Long gone by then, the Wildwood was planted with rows of sitka spruce after 1945.

After the track crosses a tarmacked C road, originally a road made by Rome, it climbs straight up to the southern ridge of the valley and plunges into the dense pinewood, disappearing into the darkness of the past.

No one travels the Long Track now, except for my family on their way home to our farm and its houses. Once a busy medieval and early modern highway, it is forgotten, a road to nowhere, or at least nowhere much. But the Long Track seems to me to be a palimpsest, a highway for our history, a route to memory and, I believe, a different means of understanding how time passes. When I walk its curving, embedded length, curling around the flanks of knowes and ridges, sunk between high earthen kerbs by the tread of centuries of people, the creak of cartwheels and the hooves of horses, I sometimes look over my shoulder to glimpse the ghosts of the past.

The Length of Fetch

At Dalmore eternities meet. On the Atlantic shore of the Isle of Lewis, a horned bay shelters an ancient graveyard. When the dead come, they are carried, their coffins borne on the shoulders of mourning men who walk the long road down to the sea. As the bearers tire, some of them stooped with age, others take turns to complete a journey all will make one day. On wild mornings, when spindrift whips off the breakers, the gravediggers wait while the black-robed minister says the parting words. As soon as the principal mourners let the cords slip through their hands and the coffin rests at last on the sandy soil, the men take the sods of tussocky grass off the spoil heap and begin to work quickly. After the handfuls have rattled over the varnished lid, they begin to shovel in the covering soil before the ocean wind can cast it away. Once the lair is full, they place the grassy clods on top and stamp them down.

Many headstones bear the same hope: *Gus am Bris an Latha*, Until the Break of Day. On some, another line is added: *agus an Teich na Sgailean*, and the Shadows Flee Away. At Dalmore, the gravediggers make sure that there will be no resurrection until God ordains it. Blowing across the wastes of the mighty Atlantic, the west wind would not disturb the dead until the last morning, the day the trumpet sounded and the dead were seen alive. Only then would dawn break over the endless horizons of eternity. Only then would time collapse on itself.

At the tips of the horned bay, cliffs rise, and to the north-west sea-stacks stand against the surge of the ocean. On the sweep of the beach below the graveyard, above the high-tide mark, beautiful, rounded grey boulders are piled up in a dense curve. Many are veined with pink and blue, all have been tumbled, rounded and wave-worn by the power of the ocean over countless millennia, as it rumbled them relentlessly towards the beach. Some seem to be Lewisian gneiss. Amongst the oldest rocks on the Earth, these strata were pushed upwards to spew through the young planet's crust as it burned and convulsed many millions of years ago. Now they lie still, quiet and dignified, their immense journey ended on Dalmore beach, settled below the rows of headstones.

When the breakers roll in off the Atlantic, they rise and fall on the beach with a whump, like the beat of the world's drum. It is loudest when the seas are mountainous, when the waves rise high, cresting white foam and crashing. Mariners and meteorologists know where these leviathans begin their lives. What they call the length of fetch is the distance a wave travels before it beats against the shore and retreats invisible in the undertow. When a westerly wind blows directly across the Atlantic from the coasts of Canada, it creates waves three thousand miles from where they make landfall, an immense length of fetch.

At Dalmore time turns to a different rhythm. It is not linear. Now, we are all children of clocks and calendars, caught up in the dance of time, the cares of tomorrow, the month's end, the next minute. The boulders of the beach of the horned bay, the whump of the breakers and the hope carved on the headstones speak not of the transit of days but of the endless renewals of millennia. By the cemetery at Dalmore, where the dead sleep in the sands of time, I began long ago, almost without knowing it, to think of my life in another way. Instead of looking at my wrist to discover the time of day, and become anxious about what was next, it occurred to me that I should look up at the sky, sense the shift of the weather, the seasons, and glory in the day, even in the moment. In a Gaelic phrase, I learned to listen to the music of the thing as it happened. It has taken many years for me to begin to shed the compelling notion that by organising time we somehow control it. We do not.

On a warmer, kinder shore stand the ruins of Hippo, a Roman town on the Mediterranean coast in the province of Numidia, modern Algeria. There a theologian struggled with notions of time. In his fourth-century *Confessions*, St Augustine wrote, '*Quid est enim tempus?*' For what is time? After years of thought and prayer, much of it concerned with the Second Coming, the End of Days and the Last Judgement, he came across a simple metaphor. In an image that prefigured modern measurement, Augustine understood time as a potter's wheel. It is an attractive notion, offering a sense of process rather than a linear progression. The wheel would stop turning when the pot was formed. And would only turn once more when another lump of clay was puddled, rolled in the potter's hands and placed on the wheel.

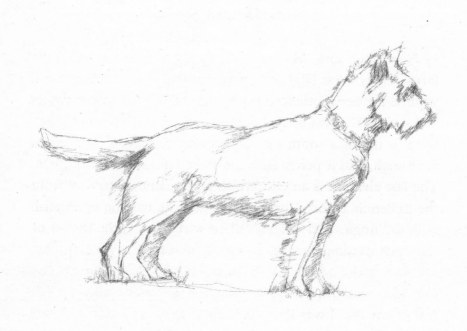

Introduction

Maidie and Me

In November 2016, Maidie came to the Henhouse, our little farm. A tiny West Highland terrier puppy, she whimpered and whined for her mother and the warmth of her little brothers and sisters. She was born and bred at Skelfhill, a remote sheep farm in the hills south-east of Hawick. Skelf is a Scots word for a shelf, and it perches on the steep flanks of Skelfhill Pen. The last element is an Old Welsh place name, a survival from the millennia before Christ's birth and the coming of English with the Angles and Saxons. When we parked in the shelter of the stone steading, I realised I was about to make a life-changing decision. From a litter of five puppies, I was about to choose one who would be a companion and a responsibility for the rest of my life. I was sixty-six then; if they are hardy and well looked after, terriers can live for fourteen or fifteen years, sometimes longer. If I was lucky, we would grow old together.

When we first came to the farm in 1994, we were given two Border collie puppies by accident. Meg and Kelsae were sisters but very different in looks and temperament. They were born at the farm on the other side of our little valley, and when my family first mooted the idea of having dogs I objected. They would be very tying. Not only would a dog need to be walked, fed and cared for, every day, we would not be able to go anywhere on holiday as a family unless we parked it in boarding kennels. In the teeth of relentless, calm and well-argued advocacy from my son, my misgivings withered and after a final discussion we drove down the Long Track to see the puppies.

My wife had said we would take one of the bitches and, with my three children, had chosen Meg. To distinguish her from the other little ones, she had a small tuft of white hair on the back of her neck and a very glossy black coat. When the litter was at last weaned and we went to pick up Meg, the farmer's wife told us that one of the other puppies had not found a home, the person who had agreed to take it having changed their mind. I capitulated immediately, and Kelsae came back to our farmhouse with her sister to begin a life of almost fifteen years at the heart of our family and safe in our family's heart. They were loved with a fierce tenderness, and as the collies grew up, so did my children. It was a life-enhancing experience.

When Meg and Kelsae died, days of many tears and immense sadness, the farmhouse seemed suddenly empty, cold and quiet. Where the collies had dozed in front of the woodburning stoves, there was just a carpet and some old sheepskins. We lasted three weeks before Lillie came. A lovely labradoodle, white-coated, cuddly, very different from the independent, keen-eyed collies, she was a snow-pup, coming to the farmhouse when the first drifts fell in the severe winter of 2009–10. They were so deep that Lillie had to jump and swim through them, only her black nose and eyes showing, like three lumps of coal, a tiny, dancing snowman. At first she seemed content, immediately claiming a place beside the crackle and glow of the fire. But when the summer at last came we sensed that Lillie was withdrawing a little, not sitting beside us in the evenings. Perhaps she needed a companion, and Freydo (a conflation of preferences: my wife called her Freya, and the children think she is like Frodo Baggins) eventually came two years after Lillie.

Another labradoodle but, as often with crossbreeds, they were very different.

Freydo and Lillie are family dogs, and very much loved. For some incoherent reason, I decided that we should also have a West Highland terrier. Jaunty, full of confidence and with great presence for a small dog, they made me smile when I saw one in the street. There is a Scots word, perjink, that captures the essence of a Westie. It means 'neatly made', but carries overtones of naughtiness.

The puppy from Skelfhill was to be my dog, my companion, and in a moment of great presumption I named her after Walter Scott's faithful Maida. Scribbling about the history of the Scottish Borders is in reality the sole link we have, certainly not literary merit or book sales. Immortalised in stone at the foot of her master under the canopy of the Scott Monument in Edinburgh, Maida was a huge Irish Wolfhound. My Maidie is not huge, but she thinks she is.

When we drove away from Skelfhill, I held my puppy in my arms, a tiny bundle wrapped in a small blanket the breeder had given me. It carried the scent of her mother and the other puppies. I remember Maidie looking up at me very uncertainly. She was so small, confused, and as we threaded our way through the hills she began to tremble, even though it was not cold.

At home the two big dogs sniffed her and seemed not to be discomfited. I kept her wrapped in the little blanket and at night we put it in her crate. For about an hour, Maidie whimpered pitifully in the dark and I had to be restrained from going downstairs to pick her up. That first night was the making of a strong bond. I was responsible for taking her away from the rough and tumble of her litter and the sweet milky scent of her mother, and I had to replace that loss of love. The first few months of toilet training were very tough and she took an age to learn, and so did I. More than once I wondered what I had done.

From the very beginning we walked, Maidie and I. Small dogs feel the cold and to see her through the bitter months of January, February and March I did something I used to mock when we lived in Edinburgh. I dressed Maidie in a tiny fleece coat. At least it was not bright red tartan, and no one saw us as we went for walks around the farm, me imploring her to poop and pee. We followed the Long Track, but also other routes, and I came to realise that my daily excursions were becoming much more than exercising a dog. I had always done jobs on the farm with the horses my wife breeds and breeds from: fencing, repairs of all kinds, taking feed to the outbye fields, woodcutting and much else. I thought all of that activity meant that I knew this place well, but in reality my appreciation was only superficial. On our daily, early morning walks in all weathers and through all the seasons, I really began to look with a much sharper focus, to observe the land, its plants, its animals and the weather that governs it all. While Maidie sniffed after long-departed rabbits, I stood, increasingly patiently, and watched, and saw things I would have missed had I been going about the business of dragging logs out of the woods or nailing broken rails back on fence posts. High in the early winter sky, I heard the plaintive honking of arrowheads of migrating geese; on the edge of the woods I could make out the faint tracks where roe deer had flitted through the long grass; and I came to love the intense scent of the sweet poplars by the East Meadow. Maidie showed me where I lived, and I became deeply interested in the life of the land and its story.

I decided to keep a daily diary. Not just an aide-memoire of events, it would also record changes in the weather, the shift of the seasons, the growth, life and death of plants and animals, and what I felt and thought about what I saw unfurl through the year. It would also be a diary of discovery. Instead of wondering about the story of the Deer Park, a low hill that dominates the north of our farm, I would use all my experience as a historian to attach some facts to vague suppositions. Instead of wondering about the transit of the constellations across the night sky, I would discover how they were understood in the calendars of the deep past, a time before a modern under-standing of time. Raised in the late-Victorian countryside, my grandmother told me once that she could find her way home through the winter darkness by starlight. Was that true?

When a house was built on the farm for my son and his family, the digger broke into an old stone drain. It had been beautifully made. Before the era of factory-made ceramic and then plastic pipes, skilled and hardy gangs of men had dug long trenches across the field behind the new house, laid a course of rounded river pebbles and then larger, flatter stones on top. It was slow, back-breaking work and I was vexed to see it smashed by the bucket of the digger, but after it rains water still trickles out of it. Who made these beautiful drains, and when?

Everyone lives in the midst of history. Cities and towns are constantly changing, their buildings replaced, roads re-routed and horizons re-formed. Where there is demolition, when an office block or an old building past its time is removed, like a tooth extraction, what the muddy, messy gap reveals is history. All towns and cities are other palimpsests, their present piled directly on top of the past, hiding stories, layers of centuries, but not obliterating all that experience in one place. Street names remember ancient purposes, a time before the internal combustion engine, supermarkets, electricity: Horsemarket, Poultry, Potterow, Nunnery Lane and innumerable Castle Streets, Abbey Streets and Quay Streets. All that is needed to convert these noisy, busy places into rich narratives is curiosity. City streets are not too dead for dreaming.

Having left Edinburgh more than twenty years ago to come with my family to live in the countryside, I took some time before I began to think about the land I walked through, its trees, plants and animals, and all of those people who walked our fields and tracks before me. But when I did, it turned out to be the beginning of an intimate and loving relationship. What I discovered fired my imagination and made the small world I inhabit enormous, a place from where I could reach back through time, across millennia, so that I could take the hand of those who lived where we live eight thousand years ago. They saw the same skies I see. They were once me and I am them.

What follows is the story of my year, but at the same time it could be any year and all years. Every place has its history, its secrets, and all that is needed to unlock and learn them is love, patience and open eyes. It is, of course, arranged into days, weeks and months, but although dates are attached, no year is. The seasons change, the winds blow, the rain falls and the sun shines, but cracks in time will continually open as the past slips through and onto the pages of the present.

January

1 January

Six hours after the year's midnight, the light of a waning crescent moon glints off the ice on the track down to the stable yard. Out early with the dogs to let them pee, we walk under an open sky glittering with starlight. Appropriately, the brightest is Sirius, the Dog Star. It shines in the southern sky long after the others have faded. Even though sunrise is more than two hours away, the eastern horizon is blue and will soon glow pale yellow. The dawn of the new year will bring sun, hope and welcome warmth.

The world is slow to wake after what we used to call Auld Year's Night in the Scotland I grew up in. Better than New Year's Eve, it reflected the year's turn, the Janus-like moment of farewell to the darkness of the old before the dawn of the new. It was a time of licensed excess, when even respectable people could take a dram or two, or three, before the bells rang out and hands were shaken, wishes made and kisses given. Those overcome by too much cheer were said to have foundered, like ships ploughing through the waves before running aground, as they made their deliberate, unsteady way to bed.

When Maidie and I reached the top of the Long Track, we looked out over our little valley and saw that it was still fast asleep. Three miles away, on a shoulder of the western hills, not far from the old Roman fort at Oakwood, the farmhouse commands a sweeping vista, looking back at us in the east. Here the Hartwoodburn begins to carve out the shape of the valley as it rises near the foot of the Long Track before forking around the wide, billiard-table-flat grass park known as the Tile Field. Only the distant lights of the kitchen windows in the hamlet of Hartwoodburn were twinkling, eight houses and a farm steading strung out along a single-track C road. On the other side of the Tile Field, on the south-facing slope below the northern ridge that fringes the valley, is our little farm, the Henhouse, and its outhouses and stables, sheltered by trees. Out to the west, the Hartwoodburn is joined by other streams flowing down off the hills before it tumbles over a waterfall called the Motte Linn and then spills into the River Ettrick.

I should pause here for a moment of definition. I have written above that we live on a small farm in the Scottish Borders, but I am not absolutely sure that we do. It is difficult to know what to call the bit of the planet we own and look after. At eighty acres, it is bigger than a smallholding or a croft – I would wince if anyone called it an estate – but it seems too small to be a farm. We do not grow any crops except grass and my wife manages about twenty-five acres of pasture to breed horses. So, I guess small farm is the least inaccurate description.

The drams of Auld Year's Night having kept many in their beds, Maidie and I walk through a silent landscape where little stirs. No car headlights swing up the Thief Road into the western hills or sweep down through the trees from Greenhill Heights. Across at Hartwoodburn Farm, where more than a hundred head of cattle are in their winter byres, it is quiet. By 7 a.m. the farmer is usually busy clattering around with his digger, lifting silage out of his pits and into the mangers for the morning feed as the cows trumpet their joy, steam rising into the cold air from the fermenting forage. But he is late and the hungry cows wait quietly.

Lindsay and I celebrated Auld Year's Night with some good wine and trusted that the New Year would safely come in without our help. With our son, his wife and our granddaughter Grace living on the farm, as well as our daughter, Beth, and her husband visiting, Christmas is much loved and much preferred as a family festival. But it was not always so in Scotland. Not until 1958 was Christmas made a public holiday; the focus was traditionally on New Year. I can remember my dad going to work when Christmas fell on a weekday. Early in the morning, before he left the house in his overalls, we were given one or two presents and a stocking with a tangerine, half a crown and some sweets, usually Toblerone, which I never liked. When Dad came home from work, we had roast chicken, a real treat. It was television that changed attitudes. Before the changes of the late 1950s we used to watch England having a party and Scotland seemed not to be invited.

2 January

All that disturbs the quiet of the early dark is the rustle of Maidie's paws as she stalks through the crisp and frosted grass, sniffing after the night trails of voles and mice. I pad up the track behind her and then turn to look east for the first glimmers of morning. The Christmas and New Year holidays are the only time of the year when no one is sure what day it is. The rituals of meals, gifts, toasts, daytime television and long walks are what give structure and not the calendar and the weekly round. It is as though ordinary time is suspended. Most significant for us is the passing of the solstice and the longest night. Very slowly the days begin to stretch and, while the worst of the winter is still to come, the light is returning.

At Windy Gates, at the top of the Long Track, the vistas in all directions are wider than they were last winter. A dense sitka spruce and Scots pine wood that belonged to my neighbour stood above the Top Track and last July huge machinery rumbled around it to begin a savage process. Like a dinosaur in the primeval forest, the harvester's saw swung on the end of an extending arm – it could fell a tall tree in less than five seconds. The scruffy sitka had been planted in 1950 and needed to come down before they were all blown down, but I was vexed to the point of tears to see the loss of the Scots pines. They marched along our track to make an avenue with the birches and geans we planted on the opposite side and they sheltered our home paddocks from the bitter north winds. Majestic, red-barked and with deep forest-green needles, those trees were my friends and I miss them.

The land changes but it never ceases to dream of the past: primitive, spiky marsh grass reclaims boggy fields, springs insist on avoiding drains and bubble up through the stones of my tracks, and the unchanging, eternal wind blows on, whistling, filling the air with the sound of history. Last September, when the trees were still in leaf, a storm burst on us and the two old sycamores left by the harvester (the felling licence was for pine only) suffered terribly, with half of their main limbs torn off. Yesterday afternoon, my son and I cut some into manageable lengths and stacked them in the wood yard to season. The heartwood was milky white and the rings so close as to be uncountable. I am glad to have the hardwood logs for our woodburners, but sad at the destruction.

The densely ranked woods of the commercial forestry planted since the Second World War are sterile, dark places where the deer find cover, but little else, plant or animal, can thrive. The regiments of sitka spruce mask the shape of the land and blank off vast areas so completely that they can seem invisible. The loss of the Scots pines by the Top Track was partly compensated by the opening up of the northern ridge. From a new vantage point, I was better able to see how our little valley might have looked to those who walked their lives here before us.

The past of this place fascinated two men I wish I had known. Bruce and Walter Mason were bakers in Selkirk, our nearest town, and since the 1920s they had been field walking in and around our valley. After spring ploughing, and ideally after a shower of heavy rain, the brothers quartered fields to see what the plough had turned up. Sometimes they picked up valuable objects glinting on the top of the rain-washed furrows: a Roman enamel brooch, silver coins and other items. But most eloquent were the flints collected by the Masons. Some came from the fields of our farm, and their discovery and dating radically shifted my sense of this place. The flint arrowheads and spear tips described a very different landscape. Ten thousand years ago the flat Tile Field was a lake, a small loch, and it seems that family bands of hunter-gatherers may have sometimes overwintered around its margins.

The gossamer traces of their shelters have long since fled. Built from the boughs or trunks of trees and floored with rushes from the lochside, these structures were entirely organic and will have melted back into the soil they came from. The south-facing bank of the shallow loch was a good place to wait out the storms, frosts and rains of the winter. On calm days fish and eels could be caught and the wildfowl that nested along the damp, reedy shore could be netted and their eggs collected in the spring. Many hunter-gatherer bands harvested hazelnuts in the autumn, roasted them and ground a paste that would keep through the hungry months. When summer came, the band who hunted around the Tile Field will have moved on to other ranges where seasonal prey ran and where the wild harvest was good.

Walter and Bruce Mason's sharp eyes and vast knowledge lit the darkness of the past. By recognising these small, razor-sharp pieces of flint, and understanding their manufacture and their uses, they reached back across ten millennia and made all that experience in one place come alive. They knew that history is like an unmade jigsaw whose pieces lie scattered about us on every side.

3 January

A vixen shrieks. Even though she is deep in the wood, waiting for her mate to come to her, her shriek echoes around the little valley. A damp winter mist has settled on the surface of the loch, Hartwood Loch, its reedy fringes and watery willow scrub surrounded by the woods where the harts flit between the shadows, the wraiths of the Deer Wood.

The vixen's piercing call stirs the sleeping family. On a bed of brackens and brittle leaves, a mother draws her son, daughter and their father closer. Though they wear buckskin tunics and pull pelts tight around them to keep out the shivering wind, their warmth comes from each other. They live their brief lives in the midst of wildness and know that the mating foxes present no threat. Set in a circle of flat hearthstones, the smoored fire is no more than dull and feeble embers, but with dry lichen and slender, barkless twigs, the mother leans over and blows the flames into life before carefully laying on more kindling. By the side of the Hartwood Loch a spiral of smoke rises in the morning sky.

The hunter-gatherer bands who sat and knapped the flints found by the Mason brothers were few. Between 8000 BC and 4500 BC the population of prehistoric Britain was likely tiny. Perhaps only ten thousand fished the lochs and rivers and hunted in the Wildwood, only one species of mammals amongst tens of thousands of bears, bison, lynxes and pine martens. What brought them north from the Ice Age refuges in the caves of southern Europe were the migrating herds. As the weather quickly warmed, wild horses, reindeer and other fauna failed to adapt fast enough; unable to shed their warm coats, they chased the cold northwards. Our ancestors were forced to follow them, so great was their dependence on their meat, pelts, sinews, hair and horns. But as the sun shone and the snow retreated to the highest mountains, the land came back to life after the millennia of ice, and the Wildwood, a temperate jungle browsed and sometimes opened up by grazing animals, carpeted the great river valleys of the Scottish Borders and reached high into the western hills. This wide area may have been home to about a hundred people, perhaps twenty or twenty-five family bands. Many of the animals of the Wildwood never saw a human being and, when they did, may not have fled before the flint-tipped spears were thrown or they were chased towards the camouflaged pits dug in forest paths.

My old friend Walter Elliot knew Bruce and Walter Mason well, and although they both are now long dead, he told me recently he still misses them. Shy and diffident men, they had walked the fields around Selkirk collecting all manner of objects, as well as papers related to the town's history, for a simple reason. They were curious. Neither was formally educated, but they shared all the attributes of the supremely well self-educated, much rarer and greater prizes than a piece of academic paper. Walter shares those singular virtues and I shall be eternally grateful to all three men for their patient work and keen observation. They have made my life in this place richer.

The vixen shrieked in the sitka spruce wood near Windy Gates when I walked out in the early dark with Maidie. Known as the Young Wood, it was planted twenty years ago after its exact replica had been harvested and carted off to the sawmill. Her mating call was only a distant echo of a wildness long vanished. Much more often the morning air carries the bleat of sheep on the western hillsides or the murmur of cattle feeding in the byre. As I made my way back to the farmhouse, I saw the lights of the shepherd's quad bike as he rode up the southern ridge to check on his ewes.

We have tamed the planet and only the elemental wildness of the oceans and the weather remain. Savage storms can destroy in moments, while long sunny evenings can send us to bed content with creation. These forces govern all.

4 January

This morning the world was stirring at last, going back to work after the holidays. Out in the early dark, amongst the ice-white scintillae of the stars, I saw the green and red port and starboard lights of five aircraft, four flying north-west to Glasgow and one due north to Edinburgh. Car headlights breasted Greenhill

Heights, others threaded their way down the Ettrick Valley and, on the C road, the flashing orange lights of the refuse lorry beginning its country run, stopping at the houses in the hamlet of Hartwoodburn. At a tight bend in the single-track road, it met the farmer and his digger on their way to feed the cows in the byres. He had to reverse a long way to find a passing place to ease the traffic jam.

As the morning light brightened, the crows roosting in the woods by the Deer Park lifted into the air. More a dawn cacophony than a chorus, four or five hundred flocked and flew south over the Tile Field. Some landed near the Long Track and began strutting through the grass, their black chests thrust out, searching the ground for morsels. There are far too many of them. If they were the size of buzzards, crows would be terrifying rather than merely annoying. Elsewhere in the sky, there was pleasing contrast. Honking gently, fourteen geese, strung out in a breaking, re-forming arrowhead, flew south-east towards the morning sun.

As Maidie and I reached the bottom of the Long Track, I looked over to the steading where the farmer was working his digger, picking up loads of steaming silage, dumping it in the mangers, reversing, swinging round and repeating the same manoeuvres until all were fed. Especially in the cold winter dark, it is long and hard work. With only one other man, he runs two farms, rearing many hundred head of cows and sheep, as well as growing barley and hay in the lower fields south of us. In the summer, contractors will come and cut his crops, and he will sell the barley and store the hay as silage for the following winter, as the day-in, day-out labour turns the wheel and the cycle of the years unfold into the future.

None of the original four houses and cottages at Hartwoodburn are occupied by farm workers. A survey done in 1858 spoke of 'a good dwelling house with an extensive court of farm offices [meaning cart sheds, barns and byres], hinds' [farm workers] cottages, gardens and a large farm attached'. It also described the last remnants of the Hartwood Loch as 'a marshy bog of considerable extent . . .'

From my grandmother, Bina Moffat, I heard stories of a very different farm life. She was born in 1890 at Cliftonhill Farm near Kelso in the evening years of high farming. Her grandfather William was first horseman, head ploughman, and in nine family houses the census enumerator counted sixty-two people living on the farm. Twenty-three were employed full time and they raised twenty-five children, the balance of this small community being the farmer's family and those too old to work.

Each man was employed on a six-month or annual fee, new terms agreed, or not, on the quarter day at the end of each period. Cash was paid not weekly but at the beginning and end of the hiring. What was even more keenly negotiated were what all knew as gains, payments in kind. At Cliftonhill, William Moffat had half an acre of potatoes, a large quantity of oatmeal for porridge, coal, a sty behind the tied cottage to rear a pig and the run of the steading, where his hens could forage and lay. When she was little more than a toddler, Bina's job was to find the eggs amongst the haystacks, which, in the long grass beside the farm tracks, was not always easy.

Cliftonhill's gains were thought especially good since they included a significant wild harvest. On dew-drenched late summer mornings, Bina and her aunts were out early to scour old pasture for mushrooms. Fields where horses grazed were thought to be particularly likely. In the later autumn, they cleared the orchard of apples, pears and Victoria plums. After the farmer had selected what he wanted, they preserved as many as possible. Down by the banks of the River Eden, they went nutting, stripping the glossy, brown hazelnuts off the bush-like trees and also picking as many wild raspberries, brambles and elders as they could find in the hedgerows before the birds got them.

Part of the bargain, but unfurnished, their cottage was small and Bina slept in the same box beds as her mother, Annie, and her aunts, Mary and Bella. On cold January nights they were each other's warmth, and when I was a little boy I slept in my grannie's bed, enveloped in her love. To avoid collapsing into the deep dent she made in the horsehair mattress, I lay on my side on the edge nearest the window. As Bina snored contentedly, I remember looking at the full moon, liking that its beams lit the room. I have never been fond of single beds.

The great French historians Marc Bloch, Lucien Lefevre and Fernand Braudel wrote of the longue durée, the persistence of similar habits of mind and action over long periods amongst communities who were born and died in the same place. The link between what Bina called the *auld life* and the hunter-gatherers who overwintered by the Hartwood Loch ten thousand years ago is not tenuous, or merely genetic. At Cliftonhill Farm, just as the hunter-gatherers had done before them, they roasted hazelnuts (Bina told me they tasted better and were easier on your teeth) and around the vanished loch I have seen places where mushrooms and abundant berries could be found.

But the precious, unconscious continuities of the longue durée are breaking down. Storm-force winds of change have blown in my lifetime, and they blow harder with every passing year. The past is being ground to pieces by instant global communication, by the rapid, ever-updating output of the internet, by blizzards of relentless novelty. We have mistaken ease of contact, convenience, consumerism and personal comfort for civilisation, and our identities, those fragile, complex characteristics that make us interesting and different, are being interpreted by algorithms and buried under snowdrifts of digital data.

Population shifts have also been dramatic. Until very recently almost everyone was directly or indirectly involved in food production, most working on the land. As late as 1800, 80 per cent of the British population lived in the countryside. Now that has shrunk to 1 per cent and there are only 150,000 farmers, the majority working alone with the occasional help of spouses and family.

The land has almost lost its people and now its memory is beginning to die.

5 January

Yesterday afternoon a drunk smashed into one of the farmhouse windows, startling me as I was reading by the fire. But, having rushed out of the porch, I could see no sign, no prone body at the foot of the wall. Under the apple trees opposite the house I could see others becoming intoxicated. After the stormy weather of late November, we had left many windfalls on the grass. Bruised, they would not have kept. Instead, they turned out to be a winter windfall for the birds – blackbirds especially. I counted fourteen this morning, gorging themselves.

As the apples begin to rot, they ferment a little, and when the blackbirds hollow them out they become tipsy. The drunk who flew into the farmhouse window was the third that day. Like the others, it recovered immediately, surviving the impact with the glass like those rubber-legged New Year revellers who collide with pavements, walls and each other, yet wake up the following morning with only a bad headache.

Warming thoughts of spring light a cloudy day. Last year, when I was recovering from shoulder surgery and had my arm in a sling, I took up one-handed gardening. Two raised beds were built in a small paddock by the burn and they were filled with well-rotted horse muck, something that is not in short supply around here. My new potatoes were sweet and splendid, smooth-skinned and a meal in themselves with melted butter, salt and pepper. When we harvested them, I lifted my two-year-old granddaughter Grace into the beds. No graip was needed because the soil and muck had become a fertile powder, and as I pulled up a shaw to reveal a string of yellow-white potatoes the wee lass squealed with delight at this buried magic.

Three weeks ago I assembled a lower raised bed from some

wooden shuttering which would become Grace's garden. We filled it with horse muck and well-tilthed soil from the many molehills on the lawns around the farmhouse. When I asked Grace what she wanted to plant in the spring, 'Potatoes! Big giant ones!' was the instant reply. In the spring we will rime out the holes with my dad's dibble, the adapted wooden shank of a shovel that is older than I am, and bury the magic once more.

More pressing is the need for logs. Like many old farmhouses, ours is draughty and the woodburning stoves need to be fed constantly in winter. With my son's help, I thought we had cut plenty of logs, but now I am not so sure. What we have is so well seasoned that even the precious hardwood burns like matchsticks. Tremendous heat is produced, but we are getting through prodigious amounts.

6 January

This is the twelfth day of Christmas, and last night the decorations came down and the tree was stripped. Today it will be dragged into the Bottom Wood to take its place alongside the withered memories of many Christmases past. We have celebrated twenty-five in the farmhouse and this year's tree was the most beautiful yet. Hung with some decorations that are older than we are, and some that are new, it made me smile each dark morning when I switched on the lights.

Now we set out on the longest, slowest month of the year. And the weather has yet to snarl. Under a thick blanket of cloud, the day begins mild and windless, with no ice forming on the track and the grass green and not frosted. Some of the day will be spent preparing for the bad weather forecast for the second half of January. Behind the stables and elsewhere there is a good deal of scrap wood that was discarded when buildings were repaired or fencing moved. All of it will go in the Wood Barn to dry off and I will pick boughs off the woodpiles to be sawn up and split.

As I rummage around the farm, making estimates and working out contingencies, I realise that I am walking in the shadow of the long past. Beside the banks of Hartwood Loch many thousands of years ago, the overwintering hunter-gatherer bands made similar calculations as they built up log piles by their shelter. Having dragged fallen trees and broken limbs from the Wildwood on the slopes of the valley, they raised them off the ground on bearers to dry and season in the summer sun.

These men, women and children were creatures of the land, part of the cycle of the natural world, and they knew a great deal about different woods, their uses, how to split them and cut them and how they burned. Of course, no record of their tree lore survives, but its echoes can still be heard. These are some verses of a very old woodburning poem that originated in Devon. It has been updated, formalised and prettified, but the embers of real knowledge glow through:

> Oak logs will warm you well,
> That are old and dry;
> Logs of pine will sweetly smell
> But the sparks will fly.
> Holly logs will burn like wax,
> You may burn them green;
> Elm logs like to smouldering flax,
> No flame to be seen.
> Ash logs smooth and grey,
> Burn them green or old,
> Buy up all that come your way,
> Worth their weight in gold.

Ten thousand years ago, fires were lit by Hartwood Loch not only for heat and cooking but also to smoke eels and fish. In harsh winters, the loch will have frozen over and stored food was all that lay between a family band and starvation. And on a cold night of hollow bellies some will have fallen asleep in each other's arms and not woken in the morning.

Archaeology has shown that our ancestors worked hard and died very young. Probably because of fatal complications in child-birth, an annual event in an age before contraception, few women's lives stretched beyond twenty and men did not live much longer. From later skeletal remains, scientists have been able to reconstruct their bodies. Because of heavy and constant work, prehistoric people were very muscular, especially in the legs. When a fallen tree needed to be dragged from the wood to the shelter, no help and no alternative form of traction was available. At least half of this tiny population suffered from degenerative spinal conditions, some of them children. They seem to have been set to work early. Most suffered from chronic arthritis, probably from living much of the year in damp and cold conditions.

Life was usually harsh and short, but not every day was difficult. Just as we do, families will have enjoyed lovely summer evenings in this little valley. Their bellies full after a good supper, they will have splashed around the loch with each other, sat on logs, basked in the late sun, looked out over the green and pleasant land, and told stories, embellished them and laughed.

7 January

Time winds down. We move slowly through the calendar, ticking off numbers, but little changes. The long night of winter seems to be endless, when the land sleeps and waits, when little stirs, and the short, grey days come and go. In cities, bright lights and bustle fight the gloom, but here there are only the pinprick points of distant farmhouse windows, steadings and stable yards to relieve what my grannie called *the dowie days of dreich Januar*.

Her memories of winters at Cliftonhill were of enveloping,

inhibiting darkness. Working days – repairing tack and machinery, feeding animals, milking and keeping hungry mice and rats at bay – were short. After the middle of the afternoon the men and working women came indoors, and Bina and her family sat around the range and its warming coal fire, the main source of light. It was a time of stories. In the black darkness, the veil between the dead and the living was thin and they shivered at more than the night chill. My grannie once told me that from the cottage window she saw a wraith. Down by the banks of the Eden Water the shape of a woman rose up out of the little river and began to glow. Wearing a long dress and with long grey hair, she became brighter and brighter as she swept up from the little river and, stretching out her arms, seemed to soar over the cottage roof and disappear into the night sky. Staring out of Bina's bedroom window at the moon as she told me this story, I was wide-eyed and cuddled in closer. I believed her. I still do.

From an early age, my grannie sewed and knitted, but the strain on her eyes meant spectacles in later life. With no electricity at Cliftonhill in the 1890s, her family depended on oil lamps and candles. Bina had clever fingers, and even when they were bent with arthritis I loved to watch her push a darning stool up inside a sock and weave together a patch so perfectly that it seemed new, intended, more than a repair. Only threading fine needles defeated her, and by the light of the sitting-room window I was glad to take the thread she had drawn through a lump of beeswax and do it for her. When she took back the needle, she smiled and called me her wee lamb.

It is a commonplace to assert that time passes more slowly in rural areas than it does in cities. Sometimes there is a sneer on the edge of remarks like 'there is no word in Gaelic that expresses the urgency of mañana'. The truth is more complicated and more interesting. Time in the countryside marches not from nine to five but with the seasons. To use the imagery of the clock face, the winter sun rises at 10 a.m. and sets at 2 p.m. and that means little work on the land can be done. But in the lengthening days of spring, summer and autumn, when the year really begins with the lambing, when hay is cut and the harvest eventually brought home, farmers and rural workers are busy all the hours of light, especially when the weather is fine.

Old time was reckoned differently. Until the Industrial Revolution created the great cities and large towns, earlier cultures measured years, months and days but came very late indeed to counting hours and minutes. The most reliable clock was the cock-crow. As a result, punctuality was a very vague concept, one that has persisted. I still say, 'I'll see you at the back of twelve', meaning sometime between twelve and twelve-thirty.

Equally, the concept of being busy puzzled country people. It had little meaning since they did the jobs that the seasons demanded, when they needed to be done. And there was usually no point in working at speed so that more could be crammed into the day. Barley could not be persuaded to grow more quickly or lambs to fatten faster. Harvest was the only time when speed might matter. If the weather was due to break, reapers might race against the rain to get the corn home dry.

8 January

I saw a shooting star this morning. Stepping out of the porch at six with Maidie, I happened to look up at the open, starry sky and a streak of brilliant white light suddenly shot from the south-west quarter to the north-east. In only two or three seconds it burst out of the darkness and disappeared. Shooting stars are not stars but pieces of rock, or even dense clouds of dust, that enter the Earth's atmosphere and are burned up in moments. Larger fragments that survive and hit the surface of the planet are meteors and, even larger and thankfully rarer, asteroids.

I stood awe-struck, replaying that moment of celestial brilliance and violence. The streak of light was much thicker than other shooting stars I had seen because the piece of rock or dust was closer to the Earth. Perhaps it might have become a meteor. Spending almost all the days of the year on our small farm, a world of only eighty acres, knowing the land, its creatures, plants and moods, I was much taken with the contrast of scale, comparing the detail of our speck of a planet with the vast, unimaginable majesty of the universe. Of course I know that the planets orbit the sun but that has a stately rhythm to it. Where the burning rock came from, travelling at thousands of miles an hour, I have no idea, but the notion of out-there, of infinite darkness, of eternity, awes me.

The sky dominated the morning. As the dawn washed over us, its colours changed in moments. A very pale blue with gold on the undersides of the horizon clouds turned to darker blue and a copper-like pink against dusty grey. And then, over the dark green of the eastern woods and the waking world below, the clouds became much larger as a high altitude breeze blew up. Their undersides began to glow a hot pink that reflected on the white bark of the silver birches that line the Top Track. Behind the clouds, the sky turned a cerulean blue. It was a glorious sight, moving, rich, one of the heart-filling joys of living here.

Cerulean blue is a colour from my past. When I was a little boy, I liked to visit a shop that sold oil paints for artists. The names of the colours on the small, silvery metal tubes seemed to speak of another world: cobalt blue, ultramarine blue, carmine red, vermilion, crimson, burnt sienna. I had no money to buy these paints and no wish to be a painter, but I loved the idea of the colours. Now I see a spectrum of epic beauty some mornings that taxes description.

We have mice. At least I hope we have mice. A few years ago we had a plague of rats. In the field to the west of the farmhouse, across the boundary fence, the farmer had planted pease as an alternative winter feed for cattle. But persistent bad weather meant he could not harvest it and the field was colonised by rats. I could see their holes below the plant stalks. They infested our stables and a huge one stripped all of the plastic insulation off the wiring in the horse lorry, causing £4,000 in damage. A pest control company with the wonderfully direct name of Surekill got rid of them, but I kept finding half-decomposed, poisoned corpses I had to bury in case the dogs got to them.

When the porch door was left open overnight, mice, I hope, had come into the relative warmth and gnawed a hole in the skirting board. I installed a sonic repellent and baited two humane traps (I didn't want dead mice and their blood in the porch; the dogs would have gone postal). But when I picked them up this morning there was nothing in them. The sonic device had worked. Hooray!

9 January

The bitter cold of the morning plumes my breath and Maidie's walk is more purposeful than usual, even brisk. The store of sniffs in the frosted grass was meagre and the little terrier only paused when a pheasant shrieked somewhere in the home paddocks. She barked loudly at the intrusion, and although the daft bird was silenced she continued, but with pauses. I realised that her barks were echoing across the little valley and that she was barking at her own bark.

Nithered is the expressive Scots word for shivering cold and even through four layers I could feel the chill penetrate. The language I was raised in is full of expressive words and phrases that describe the weather, the land, and its flora and fauna. And it does so with greater precision than modern, urban English. My dog and I walked out on a glaisterie morning, one where the puddles and wet places had a thin covering of ice, one that would break easily. Smirr is a fine rain and a slounge is a downpour. A plouter can be the result, a wade through mud.

Like Gaelic, Scots is full of subtle shades of meaning in its description of the natural world. It was important to be lexically tight because the precise colour of cows, the state of the weather and what it might become, and the look of the land all mattered. Before the enclosures of the seventeenth and eighteenth centuries, animals sometimes strayed and if it was a mallachie stirk that was lost, a steer whose coat was a pale, milky colour, then the search narrowed down and ownership was clearer.

Rural Scots is dying because not enough people speak it and few of those who do are passing it on. I bear a shared guilt for that loss. When I went to university, in those far-off days when free education was a right and not an expensive commodity, I was forced to adopt a version of standard Scots English and abandon the language of my younger days. My children speak a version of English that owes more to the Americanisms of the internet than my native Scots. Not just a means of expression will be lost, so will a way of seeing the world, understanding better a way of life.

We will have welcome sun today, but even so the woodburners have been lit. Lindsay and I will both be out this morning but our three dogs, the firedogs, will bask in their glow.

10 January

There was no warning. At the corner of the Top Track, where a young wood borders the wrack of the barren area where sitka spruce and Scots pine once stood, Maidie launched herself on an urgent scent, a fresh trail that she followed frantically, her nose an inch off the ground. And then she stood still and barked at the edge of the trees. Out of the darkness a roe deer stag erupted. No more than six feet from me and the startled dog, the buck raced past us along the Top Track and jumped a high gate before disappearing into the East Meadow.

I was so taken aback, shocked even, that I heard myself catching my breath before Maidie almost pulled me off my feet in an attempted pursuit. Quite how a West Highland terrier no more than fourteen inches high and weighing less than a stone would bring down an adult roe deer stag was lost in the red mist of the moment. I was glad I had decided always to take her out on a lead. If Maidie had been off it and chased the deer, I would never have seen her again.

The early dark of the winter is a bridge between worlds. While we all sleep snug in our beds, the world of the night wakes. Badgers climb up out of the huge sett on the edge of the Deer Park and prowl the woods and the margins of the fields and tracks for prey. These little grey bears can be vicious. One of the reasons we now rarely see hedgehogs is because the badgers hunt them. Their classic defence of curling into a ball of sharp quills is not effective. The badgers simply roll the hedgehogs over and tear at their soft bellies with their long claws.

Out late in the winter darkness with a torch to investigate a burst of whinnying and banging down at the stable yard, the beam caught sight of the culprit. The red retina of a big dog fox flickered before he disappeared into the thick hedge behind the looseboxes. Last year I came across the corpse of a young vixen in the East Meadow field where the brood mares live out. She lay on her back, in the open, close to the shelter where the mares stand in the worst of the weather, but there were no attack marks on her of any sort. I think she may have stalked too close to one of the mares in the night and been kicked in the head so hard that the vixen haemorrhaged. I picked up the corpse by the hind legs and saw that rigor had not yet set in, meaning she had died during the night that had just gone. The red of her coat and white down of her belly fur was beautiful, but the day after I noticed that the vixen had been dragged away out of the wood where I had put her.

We fear the darkness and its creatures for ancient and good reasons. The world of the night can be vicious, elemental.

11 January

I have ten summers left, if I am lucky and remain in reasonable health. Next June I shall be sixty-nine and it occurred to me that I thought of the rapidly contracting future not in numbers so much as seasons. Older cultures counted years in winters because it was a time of rest, of relative inactivity. Early medieval Scandinavia and Anglo-Saxon England reckoned time in that way. Numbers apparently do not matter much to the Nuer people of South Sudan. Their sense of the past is expressed in a memory of events: the year of the flood, the year of the bountiful harvest, the year of the drought, and so on. How they remember a year when nothing much happened, I am not sure.

12 January

Silhouetted stark against the morning sky, our old oak watches over us. The tallest tree, it stands on a small hummock of rising ground by the burn that runs behind the stables. It has drunk deep for centuries from the pure water that runs at its roots and may be four or five hundred years old. With no shelter from the south-westerlies that whistle across the flat Tile Field, its branches have hesitated, growing in jerky zigzags, stubby and short so as to give the wind no purchase. The old oak's immensely thick trunk is covered with large burrs, patches of knobbly wood that cover old wounds.

This venerable tree was in its first flush of youth, standing on the banks of the Tile Field when it was still a loch and when James VI of Scotland was travelling south in triumph to be crowned in 1603 in London as James I of Great Britain and Ireland. Its ancestors colonised the Tweed Valley in the centuries between 7000 and 6500 BC. In oak years, that great arc of time might be counted as the span of twenty generations. In human years, that may be about four hundred generations. Old oaks are silent witnesses to history.

When the Mason brothers and Walter Elliot walked the freshly ploughed fields under the budding branches of our oak's near neighbours, they picked up flints that spoke of immense and now impossible journeys. The origins of these lost or discarded arrow-heads, spear tips and scraper edges are discovered by typology. Different and sometimes far distant prehistoric communities knapped flints in different ways and from different sorts of rock nodules. At the Rink, a nearby farm where the Ettrick flows into the Tweed, thousands of flints and fragments have been found. River meetings were like the junctions of main roads because most longer journeys were made in small boats. On the river terrace below the Rink, the Masons and Walter discovered flints that had made a very long journey. They had come from what is now Denmark.

Stretches of that journey were probably made on foot. Until as late as 4000 BC much of the North Sea was dry land. During the last Ice Age, a vast ice cap had pressed down hard on Scandinavia and the Arctic and had caused the Earth's crust to the south to bulge upwards. What archaeologists have called Doggerland (after the Dogger Bank, once a range of hills) was for four or five millennia a vast subcontinent patterned by hundreds of river valleys, lakes and hills. Undersea maps made for the oil industry have revealed a drowned landscape that was home to many communities of hunter-gatherers. They rustled the leaves of the lost forests, gathered their bounty and fished its rivers and lakes. And some communities, who originated in what is now Denmark, probably over the span of several generations moved across the vast subcontinent to arrive at the Rink and knap their distinctive flints on the river terrace. Or perhaps those skills were simply transmitted across Doggerland. Or maybe the flints themselves passed through several pairs of prehistoric hands.

The Doggerland communities may also have been prehistory's

first architects. On the Northumberland coast archaeologists have found traces of a remarkable structure. Dated to between 7700 and 7600 BC, all that remains is a series of postholes. Their size, depth and placing allowed a hunter-gatherer shelter to be reconstructed. Using the trunks or the larger boughs of trees, these people built what looked like a massive version of a tipi or wigwam. The angle of the postholes showed that the trunks were canted inwards and secured at the crown before the gaps were stopped up by turf and other infill. The ground plan of the wooden wigwam was an oval shape with two stone hearths and large enough to accommodate six to eight people. Found on the western edges of Doggerland, this building was more than a shelter; it spoke of permanence. Perhaps it was a winter base camp from where summer hunting expeditions set out into the interior and the shadows of the Wildwood.

13 January

For much of the night the wind growled around the house, rumbling and gusting, rattling the slates. When I woke, there was a real chill and as I switched on the downstairs lights chaos greeted me. The double front doors were wide to the wall, thrown open by the high winds, and papers, magazines, books and a box of Christmas decorations had been strewn over the floor. But much worse, Lillie had gone.

Our lovely dog was nowhere to be found. All sorts of dire disasters – dog theft is now distressingly common – raced through my head as I shoved the doors shut and bolted them. Having roused my wife, I went outside with a torch, calling her name. And found her very quickly. Thank goodness. Following her Labrador instincts, she had found food, the windfall apples that had intoxicated the blackbirds.

Of all the elements, I fear the wind most – more than ice, snow or heavy rain. Storm-force winds can and have been immensely destructive here, and wind-driven winter rain can reduce animals close to exhaustion, even death, if it penetrates their coats and there is little respite. The horses hate it.

All of the older houses in our valley were built with their backs to the prevailing southwesterlies. Front doors and most windows were let into the eastern façades. When we renovated and extended the farmhouse, the original building had only one window looking to the west, and it was needed to light the staircase. I have seen a wind rose for here. It shows the directions of the prevailing winds that blow in the western Scottish Borders. Like a compass rose, it has the sixteen cardinal points marked, from north to north-north-east all the way round to north-north-west, and it expresses the frequency of the winds by extending the length of the sixteen spokes. By far the longest is the southwesterly.

The same wind has blown across this landscape for millennia and its direction is detectable even on flat calm days. It moulds the shape of the trees, especially those on the fringes of open areas like the Hartwood Loch. On the windward side, fewer branches grow and they are almost always shorter and less abundant than those on the leeward side.

Two more prehistoric wooden wigwams have been discovered recently, one on the East Lothian coast and the other near South Queensferry, when the approach roads for the new Forth road bridge were excavated. Their sites show an ancient awareness of the power of the winds, for the interiors were sunken, dug down into the ground, and the tree trunks raised in dips in the landscape that were more sheltered. Their conical shape was sturdy because it was much heavier at the base and the rounded sides let the winds slip past it rather than slam into any flat surfaces. More of these extraordinary structures wait to be discovered.

If the hunter-gatherers from the Hartwood Loch had built a wooden wigwam, they would have avoided its banks, where our farmhouse sits squarely in the teeth of the wind, protected only by the trees we have planted around it. Beyond Windy Gates, and on the edge of another, much more modern loch that was made larger when the nearby Haining House was remodelled and landscaped, is our East Meadow. It slopes northwards down to a little burn that has no name on the map. No more than a foot or two wide, this glinting, tinkling spring rises near an old quarry and runs for only two hundred yards. Its waters are crystal pure and our fencing incorporates it into the fields so that the horses can drink without the need for a trough. But, as important, the foot of the East Meadow is a calm, sheltered place. Even when the wind rages and snarls out of the south-west, it is a relief to walk down the track into the sudden quiet of that dip. If we have the remains of a wooden wigwam, they will be found there.

14 January

There seemed to be no light in the east, and yet it was not black-dark. Somewhere behind the thick, woolly clouds, the sun was rising in the clear blue air. Walking down the Long Track, all I could see of its wide vistas were pale shadows, darker for the woods and the heads of the hills. After the wind-blown drama of yesterday, the land was quiet. Even the crows roosting in the trees by the Deer Park were dozing.

One of the joys of the BBC Scotland website is its photographs. Sent in by members of the public, they capture, often very beautifully, the moods of the land and the cities; Highland glens contrast with the busy streets of Glasgow. At this time of year, the photographers' instinct is to seek the light. There were several stunning sunrises over lochs and one glorious shot of the meandering River Forth as it wound its way to the Firth and the sea. The cities were photographed in the late afternoon or early evening, their street, shop and hotel lights a kaleidoscope. Only one photograph was taken on a gloomy, misty, frosty morning, a view up a long ride in a sitka spruce forest that looked spooky, spiky and very graphic.

Just as animals and we do, the photographers looked for the sun in the depths of January. In the long dark, a time of rest when it was normal for people to stay indoors and pull up the covers in their warm beds, these were welcome windows on a Scotland that also waits to wake. So if I am trying to rediscover these ancient rhythms, why am I trudging through the early morning and not staying abed or pottering in the welcoming lights of the kitchen, making toast?

We all tell ourselves stories about ourselves, but only some of them are true. One of mine is that I am a morning person, the earlier the better. From the age of eleven, I did a milk round that had me out of the house at 6 a.m., and I liked it, most days. My mother said it was good to 'be up and doing' when she made me my porridge. In adult life, there was also a competitive element. When I was Director of Programmes at Scottish Television, I was first into the office in Glasgow even though I had to drive from our home in Edinburgh.

But I think there might be another, more recent story, simpler and less about long-term habits or conditioning. A close friend of my parents was a dairy farmer, and after a convivial Saturday night in the pub he always left promptly with the same phrase, 'the kye'll no' wait'. The cows will not wait. He would have to be up and doing early down at the milking parlour because if he did not bring the kye in, their udders would be bursting. He got out of his bed – every morning in life – because if he did not, animals would suffer. In a much more domestic mode, one of the reasons I get up is because the dogs need to be taken out to pee, be fed and then walked, and my brain, such as it is, needs to begin working. I have no choice.

Perhaps it is these daily, weekly rhythms – and they begin with routine, tasks done automatically and always in the same order – on the farm that make me less and less willing to leave it. A day trip to Edinburgh, or at a stretch to London, is all I can manage. I don't like staying away and I would never consider taking a holiday. On those rare occasions when I am away at an equestrian event, I still get up early, always first for breakfast, usually after a walk, without Maidie.

15 January

My neighbour has moved about fifty of his pregnant ewes down to the grass park on the eastern side of the Long Track. It cants to the south to catch any sun there is and after the mild weather since before Christmas the park is green. Grass grows at plus five degrees and some days have seen unseasonal temperatures of plus eight or nine. Lambing begins in late March here, or when the weather has turned. Up at the steading at Brownmoor, on a ridge only a mile or so to the west, the heat lamps in the lambing shed will burn all night over the straw-covered pens until the end of April. When they are lit and the first ewes come in, the shed becomes a beacon leading us out of the winter darkness.

Far beyond the farm, out there, in the country and across the world, bewildering chaos, chronic uncertainty and deep divisions are causing the post-war consensus to crack. Dramatic fissures between countries and within countries are widening as fascism rears its ugly head once more in the Western democracies. Demagogues are elected by voters who crave their simple, brutal solutions. Immigrants are targeted, social services shrink and basic human decencies are no longer to be taken for granted in public discourse.

As I read these depressing headlines, I think back to my grannie and her young life at Cliftonhill. With the gains he had negotiated, William Moffat knew that his family would be safe and not starve, whatever political mayhem happened out in the wider world. The generations might be coming full circle. Our farm could quickly be converted to produce much of the food we need if the economy paralyses, distribution crashes and there are shortages in the shops. All of the horses could be turned out in the Deer Park to forage for themselves in the spring, summer and autumn, while in the East Meadow enough hay could be cut to see them through the winter. Their muck would fertilise the hayfields as we moved them around. In the three-acre home paddock by the farmhouse, a large, south-facing and sheltered garden could be laid out and cultivated quickly. In the stables we could keep a cow and pigs, and by shutting all the doors mushrooms, rhubarb and other plants that like the dark could be cultured. This is not a pipe-dream but an insurance policy and one of the reasons I stretched myself to buy as much land as I could all those years ago. Instead of stockpiling tins of beans and tuna, I will buy lots of packets of seeds. In an old conservatory in the farmhouse I grow tomatoes and know that it is an excellent place to propagate plants – and grow the possibility of a future.

16 January

A pale dawn edges across the land and the ewes in the grass park are lying down in a tight pack, perhaps expecting rain. There is a smirr on the breeze at Windy Gates. Some mornings I look out to the east and think of Lindisfarne, the Holy Island of Cuthbert, Aidan and the saints who made the glorious gospels. I spent some time there last autumn, but I would find winter difficult on the shores of the cold North Sea, even more so in the long past. When the winds whipped across the little island, the old monks shivered in their cells, warmed only by the fierce fire of their faith and the certainty of God's undying love.

Few modern Christians would suffer so much in a ceaseless quest for eternal life, for the keys of the kingdom of Heaven, but at least we have some understanding of their beliefs and how they saw Creation. What those who hunted the Wildwood and fished the Hartwood Loch believed, if anything, is much more difficult to know. Only gossamer wisps hint at their spirituality. They carved little anthropomorphic figurines that might have been used as fertility amulets. Or perhaps they were talismans for women who hoped for easy, uncomplicated birthing. And at the end of their short lives, later communities of hunter-gatherers buried bodies smeared with red ochre, perhaps to signify blood. The clay from the Tile Field, and our fields, is red.

Long before they came north to what is now Scotland, our ancestors created cathedrals to a forgotten religion. In the Ice Age refuges on either side of the Pyrenees thirty thousand years ago, they began to use the colours of ochre to paint. At Lascaux, Chauvet and Altamira in Spain, and scores of other caves, they painted herds of galloping wild horses, a charge of the huge wild cattle called aurochs, lions, bison and many of the other fauna of the millennia of ice. Often in deep darkness lit only by fires and pine resin torches our ancestors brought to life an astonishingly beautiful menagerie of what they saw out in the frozen landscape and when the spring migrations at last began.

Why did they do this? No practical purposes suggest themselves. This celebration, perhaps veneration of the natural world seems to have had a purely spiritual meaning. The Native Americans of the Great Plains revered the buffalo, the animal their culture depended on, by dancing in imitation of the great beasts. It may be that the painted caves were indeed temples where ceremonies were enacted as fires flickered and the shadows of the hunters danced on the walls where the aurochs thundered and the wild horses stampeded. Perhaps they danced to bring them back each spring, as the animals migrated to the summer grasslands.

If our ancestors did worship the natural world upon which they depended, then perhaps we should learn from them. The climate is changing quickly as a direct result of our actions, and we need to retrieve some of their ancient reverence.

17 January

There is an elemental purity in the penetrating chill of a deep winter wind. Out in the open, no sheltering trees blunt its icy edge and it is essential to keep moving, to plunge hands into pockets to keep fingers flexible. I am in the wrack of the Top Wood, cutting logs before the weather begins to shut down. My supply is dwindling fast as I fill and refill two woodburners to keep the farmhouse snug. After a struggle with the ripcord and the choke flooding the carburetor, the chainsaw at last rips into action. Such a dangerous machine, but it works so fast and cleanly. Half an hour, one fill of the small fuel tank, and I will have enough for two weeks.

The snow is coming. Even though the morning sky is clear, the wind has backed to the north-west and forecasters are predicting falls over high ground. At six hundred feet, we are on the margin. In winters past, I have seen a light dusting on the hill of the Deer Park and nothing down at the farmhouse.

18 January

The field is fringed with one of the few scraps of old woodland in our valley. As I walked down the Long Track, peering through the early dark, I heard whispers of stories from the long past.

Near the stand of birches and alders, the Mason brothers found a trace of prehistory that expanded the horizons of our little valley far beyond the western hills and exploded the notion of an isolated, primitive community. The hunter-gatherer band that fished the Hartwood Loch and stalked the deer in the Wildwood knew of a world far beyond their familiar ranges. In rough ground ploughed around the old trees for the planting of yet more sitka spruce, the sharp-eyed Masons found a small object with a dark lustre sitting on top of the fresh furrows. It was a beautifully knapped flint arrowhead made in the shape of a small leaf, its faceted edges still razor-sharp. But it was the distinctive nature of this flint that made the find so eloquent.

It had been very skilfully fashioned from a hard, igneous stone known as porcellanite and that particular kind is found only in one place. Across the North Channel in the beautiful Antrim Glens of Ulster rises a mountain whose name remembers an ancient craft. Taobh Builleach is the mountain of blows, or strikes, the quarry mountain. Roughed-out nodules of porcellanite were hacked out of its side (*an taobh*) and exported all over Britain in the centuries after 5000 BC. Axe-heads were knapped to fine edges and many arrowheads made lethal by the stone-working skills of our ancestors.

So much porcellanite was hacked out of the outcrop in the Antrim Glens that there must have been an organised distribution network, places where travellers arrived with flint that they could barter for other valuables. The arrowhead found by the Masons on the far, southern side of Hartwood Loch had probably passed through several pairs of hands before it was hefted to the straight shaft of an arrow. What this find speaks of is regular contact, the exchange of news, a mutual language, social links and an under-standing of a much wider world. A prehistoric economy was functioning in our little valley.

What also drove small family bands to be in touch with others were some fundamental needs. As early as twelve or thirteen, children had to find marriage partners so that they could begin their own band, and that meant contact with a widening circle of neighbours and perhaps sufficiently distant relatives, the further flung the better.

19 January

The snow is teasing us, toying with us. There was a sugar dusting last night and some ice had formed by morning. It is a grey, cloud-filled day but not, I think, cold enough to snow. I think.

When I drove down the Long Track, I passed a metal detectorist. I hope he finds something and I hope we find out about it if he does. I was on my way to see another old friend, someone who might have been able to advise him.

In a comfortable cafe attached to a shop stocked by one of Scotland's foremost weavers of tartan, surrounded by the chatter of friends exchanging news and elegant displays of kilts, plaids, Royal Stewart tartan scarves, shawls, dirks and Argyll socks, Walter Elliot made me a gift of my history. Out of a supermarket plastic bag he brought treasure that had lain buried and hidden under the fields of our farm for thousands of years. When the grass parks on the northern banks of the Hartwood Loch had been ploughed, and rain had washed the crests of the furrows, Bruce and Walter Mason had gone fieldwalking. Hoping the slanting spring sun would catch an object of ancient lustre, they picked up fragments of startling beauty. The Masons had given these to Walter Elliot, and in an act of extraordinary generosity he gave them to me. 'I'm in my eighties now and I need to pass things on to people who know what they are.'

The bustle of the cafe seemed to still as Walter took out an old, clear plastic bag full of flint tools that had been worked and made by people who lived where our farm is around six thousand years ago. From the unpromising debris, he pushed out two exquisitely made arrowheads, not much larger than a five-pence piece, that had been knapped with a deadly delicacy to razor-sharpness. One was leaf-shaped and the patient, perfectly weighted blows with a knapstone could still be made out. The other was tanged, formed like a tiny Christmas tree with a thick trunk. Both arrowheads had been fired, for each had traces of the birch bark resin that had fixed them to their wooden shafts. Perhaps the bowmen missed their targets and lost them in the tangle of the Wildwood.

Walter brought out two sharp edges that had been used like knives to skin an animal, a beautifully made scraper for removing the fat from the inside of pelts, and about forty microliths that had been set into arrow shafts to prevent them falling out once they had pierced the flank of a prey animal. This would make it more likely that it would bleed to death from the wound as it ran. Found on a farm about a mile to the east, there was a heavy hammerstone. About the size of a large potato, it had been knapped so that a thumb fitted into an indentation on one side and fingers on the other. There were clear signs of wear on its surface.

Perhaps most startling were the relics of prehistoric manufacturing. The Masons had picked up two stone spindle whorls in our fields. Like small doughnuts, they were almost perfectly round, with holes bored through the centre. When making yarn for weaving, women attached the whorls as a weight to the bottom of their spindle sticks so that they would spin faster and more steadily as they formed strands of yarn with their fingers from skeins of wool. The slightly smaller one shows how the hole was bored with a bow drill, a bit of harder stone made to grind as someone spun it, so that it turned quickly as they sawed back and forth with an instrument like a bow for a violin. Remarkably, the other whorl had been decorated. Someone had scored straight lines radiating from the central hole so that they looked like the rays of the sun. Not only weapons and objects used in manufacture had been made on our farm six thousand years ago, art had been produced too.

Handling all of these objects on my desk as I write this is extraordinary, moving. I feel tears prickle at the thought and the tangible evidence of all that experience in one place, a sense of closeness to these people, in space if not in time. And I am grateful beyond words to Walter for passing on these finds. He knew his gift would make the deep past of this place come alive for me.

I have long thought of history as something personal, not remote or abstract but as events and processes that happened to people like me, to anyone. Much more than the procession of kings, queens, generals and armies through the landscape, history should be about attempting to understand something of the lives of those uncounted and unnamed generations who lived under these big skies before us, how they saw the world, what they believed, hoped for and feared.

This collection of objects suddenly made the story of this place even more intimate. I could imagine a hunter preparing his equipment, sitting in the bright sunshine by the loch so that he could see better the detail of what he was doing, knapping gently at flint flakes to make his delicate arrowhead, tutting with exasperation as one splintered, throwing it aside, holding another up to the light to assess the symmetry, its natural shape, how it would flake. On another bright day, perhaps minding her children, walking with them around the hurdle fences of their little fields, a mother spun her spindle to make the yarn for her family's woollen clothes. And then the whorl was lost, the arrow with the leaf-shaped head fired and never retrieved. They lay hidden in the earth for many millennia, waiting for Walter and Bruce Mason to find them, and through them remember those lost and forgotten lives.

I am amazed at the series of unlikely coincidences that concluded with Walter Elliot's thoughtful kindness that brought the scrapers, knives, arrowheads and whorls back where they belong.

When my family and I first came up the Long Track to the ruined cottage and the acre of land around it that would grow into our farm, we felt an intense sense that this place had atmosphere, had history. Hidden beneath the tumbledown walls and the tangle of willowherb choking the garden were stories, perhaps even spirits, the ghosts of an immense past. That is no exaggeration. The summer after we had bought the roofless old cottage and its ruined stables, we often drove down from Edinburgh on a sunny Sunday just to be here. And when the time came to return to the city, we started the car reluctantly. Swirling around the whinstone walls was a powerful sense of belonging, of having come home. From Walter's gift, I now know that at least three hundred generations had called our farm home, and since then they have whispered their stories to whomever stops to listen.

20 January

Morning mist shrinks the world dramatically. I can see no further than a radius of fifty feet but out of the echoing mirk come the sounds of activity. My neighbour clatters about the steading, feeding his cows, who trumpet their appreciation of the sweet, steaming silage. Like a squadron of miniature Battle of Britain spitfires flying out of the clouds, a flock of cawing crows suddenly appears and disappears, and invisible in the grass park a ewe coughs.

At Windy Gates stands the spirit of defiant renewal. In a grey, brittle stump that survived from the hardwoods that predated the planting of sitka in 1950, a Scots pine seedling has grown. Somehow its roots have threaded their way down through the dead wood to find the earth and some moisture and sustenance. Now three feet tall, its existence is testament to life and how it finds a way. Perhaps deep in the old stump there is a memory of all its long life of spring leaves and summers past before the sawyers came nearly seventy years ago.

21 January

Last night's full moon, what is known in January as a wolf moon, shone so brightly that it eclipsed all of the stars in the southern sky and lit the land so clearly that deep moon shadows were cast. They were the darkest places in a landscape of many shades of grey. Five hundred years ago it was also the landscape of larceny. Bands of horse-riding bandits known as Border Reivers raided cattle on clear nights like this. In winter, all of a farmer's cows were herded off the hills and corralled in the inbye fields around the steading so that they could be fed. That made them much easier to steal, and when there was a full moon the reivers' ponies – surefooted, shaggy little creatures – could find their way over the rough ground. Raiding continued into the early seventeenth century, long enough for its memory to survive in names like the Thief Road, a hill trail that became a tarmacked back road over the hills to the west and passes close to Hartwoodmyres.

A red dawn rose this morning, the light refracting through bands of horizontal clouds as though they were the slats of celestial Venetian blinds. At minus seven, it was the coldest morning of the winter so far, and although the sun streamed through our south-facing windows I lit both woodburners.

22 January

My mum and dad grew up in overcrowded tenements in Kelso and Hawick, sharing rooms and beds with many siblings and relatives. In those draughty, ill-maintained old buildings winters were harsh and the sole sources of heat were a coal fire and a kitchen range. Some had a back boiler behind the fire for hot water, but my grannie had to heat everything on the range. That meant no baths for months, only a shivering, hurried wash with a face cloth and soap.

By comparison, our council house in Kelso was a paradise. It had three bedrooms, a bathroom with hot running water, an electric cooker, and a coal fire in the front room. Everywhere else was unheated and in the winter ice crystals formed on most windows. Walking barefoot on the linoleum was like crossing a frozen loch. But there was another coal fire in the bedroom directly above the front room and it glows still in my memory. In the depths of the severe winter of 1962–3, when snow lay on the ground from early December until March, I was ill, with mumps, I think, and I spent a few days in that bedroom.

Mostly I slept, and the short days and long nights seemed to merge. Under the comforting weight of sheets, thick woollen blankets, a blue bedspread and a quilt I dreamed of strange things. The mirrored wardrobe grew so large that it filled the bedroom, crushing me against the door. The window fell out, smashing into the garden below, and I was sucked out of it up into the dark night sky, spinning in space, our house growing smaller and smaller below me.

When I felt better, my mum brought up an old radio and showed me how to tune it to the stations. A tiny light went on behind a vertical list of their names, and as I moved the bar up and down, accompanied by gurgling, crackling and whooshing noises, I discovered a world beyond the dark winter nights, a geography far beyond the familiarities of the Tweed Valley. Because they paid 2/6d a week to dream of great riches, my mum and dad did the football pools. Thinking every week that this must be their week, they listened carefully to the Saturday results programme at 5 p.m. I had never before heard of Brechin, Montrose or Forfar, Stranraer, Ayr or Dumbarton, but I looked them up in a school atlas. I had no idea where Queen of the South, St Mirren or St Johnstone were, but I assumed that Third Lanark played in Lanark (they didn't – their ground at Cathkin Park was in Glasgow).

I must still have been feverish and unable to sleep with the painful swellings under my ears because I remember the first time I heard the sonorous, slow tone of the shipping forecast: Malin Head, Forties, German Bight (what is a Bight?) and Biscay. The weather information fascinated me, as an easterly gale force 8 increasing severe gale force was followed by mystery: Low Iceland 973 slow moving, filling 992 by midnight tonight. As I twiddled up and down the listed stations, I wondered where they were in the world – Hilversum, Athlone, Lille and Luxembourg – and listened for a few moments to broadcasts in languages I did not understand.

Far away, thinking about those secret nights long ago as I wandered up the track with Maidie, I failed to notice sheet ice on the surface and slid for an alarming moment like an incompetent skater. Overnight rain or sleet had flash-frozen and I had to walk on the grass. The wee dog stopped and looked at me as though I was a lesser being. I saw snow on the southern ridge and on the western hills behind Hartwoodmyres. Like a white tide, it is edging closer.

23 January

January is beginning to hurry. At last the calendar seems to be shifting and only a week remains of the longest month of the year. So far we have escaped without meteorological mishap: no snow has fallen and very little rain. This morning's cloudless sky promises another day to tick off. The wolf moon is waning in the west but its brilliance still lights the monochrome land. It is very cold – the gauge shows minus thirteen – but there is no wind to chill the bones. For an hour or so, white light from the west will slowly give way to yellow light from the east and the warmth that comes with it. It is a magical time of day.

24 January

Last night the snow came. The light in our bedroom told me that the land would be white. So that I could safely take the dogs out at 6 a.m. I pulled on boots with ice crampons attached, an essential precaution. At fifteen stone I fall hard, and the bumps, bruises and breaks linger longer. When I looked up from the snow and ice underfoot, waiting for the dogs to finish sniffing, the eastern sky was glittering, a silver dawn rising. The day looked as though it had been borne across the North Sea in the arms of angels.

The snow had made white roads through the grey, frosted grass and behind us my footprints were interwoven with Maidie's tracks as she skittered and played, her white coat a camouflage except when her coal-black eyes and nose turned towards me.

Slowly the silver dawn turned to a rich red gold as the rising sun caught the undersides of the clouds, radiating, reaching far to the west and the setting moon. The gold glowed off the snow and made the dieback grass in the Top Wood look like ripening corn. Ten thousand dawns have been welcomed from where I stood but few can have arrived on such trails of glory. Mornings such as this make Heaven seem more than a metaphor.

26 January

Out on a blustery, damp morning to check on the horses who live out, I noticed all of them nibbling on the bitter winter grass, ignoring the forage heaped in their round feeders. There will have been little nourishment in the grass but they still preferred it. All of the snow had gone, the day was mild and the animals seemed content. Four of the horses are very old, one over thirty, and while they creak and probably have arthritic pain, they have the substantial consolation of living in an eternal present. Behavioural scientists believe that while horses have memory, and that is what makes them trainable, their overwhelming focus is on the here and now. Our Old Boys out in the East Meadow do not worry about the future and have forgotten almost all of the past.

Last night was magical, warming. For the first time in eight months all of our children and their spouses and partners were together to celebrate my younger daughter's birthday. While I cooked what my adult children still call a birthday tea and others dished out drinks, the dogs and my granddaughter entertained everyone, basking in the love that wrapped around them. It was wonderful on a winter's night to hear the house ringing with laughter.

27 January

> *'Let the weight of the axe do it.' So that the boy understood clearly,*
> *the man slowly swung the axe behind his shoulder in a short arc*
> *and turned a little as the blade bit into the birch tree. 'And then*
> *you make a mouth,' and he undercut the first blow so that a wedge*
> *of the white wood fell out. Once he had chopped several wedges,*
> *the man laid down the axe and stood with his back against the*
> *mouth and pointed. 'That is the way the tree will fall. Now you*
> *need to cut the opposite side of the trunk.'*

It is more than fifty years since an old forester showed me how to fell a tree, and it was advice I had cause to remember when my son and I cut down a wind-damaged Scots pine that threatened to fall across our track. Although I used a chainsaw to cut the mouth (having waited for a windless day) on the side opposite the direction I wanted the tree to fall, the principles had not changed. And they did not change for millennia. The old forester's words echoed across sixty centuries of men and women working in the woods of the world.

Close to our farm, and especially on south-facing slopes near water, Walter Elliot has picked up or collected forty-eight prehistoric axe-heads. None of them had been knapped from local stone and most came from a place of great mystery and majesty. The jagged ridge of the Langdale Pikes in the Lake District was the quarry that produced most of the axe-heads Walter found. High up near the summit ridge, in difficult and dangerously precipitate places, miners hacked out nodules of tuff, a very hard volcanic rock. Even though there were much more accessible and easily worked deposits on the lower slopes, they chose to climb much higher to swing their antler picks. No practical reason for this can be deduced.

When the evening sun falls behind the Langdale Pikes, it backlights them black against the sky. Majestic, dramatic, the ridge may have seemed to our ancestors to be closer to the gods. Perhaps they risked the dangers of the sky quarries because they believed the axe-heads they carried down the mountainside would be blessed in some way. And perhaps that mattered because when the prehistoric foresters swung their axes they were doing the work of the gods and changing the world.

Sometime around 4000 BC the Wildwood began to be cleared. Trees were felled and scrub and bush burned back to open up the landscape. The greatest revolution in human history was underway. Farmers had come to Britain. The conventional historical record is silent about the beginning of farming, the sowing of crops and the domestication of animals. There are no recorded events, only very approximate dates, no named individuals, no battles and no conventional archaeology. Evidence for this revolution can only be found on the land and in our bodies.

Pollen analysis undertaken at Blackpool Moss, about two miles west of our farm, showed that the trees of the Wildwood – oak, elm, birch and hazel – were being cut down from about 4000 BC onwards. The Mason brothers began to find concentrated scatters of debris from flint knapping, including fragments of axe-heads, on south-facing slopes that often led down to boggy, flat areas or lochans. The Wildwood was being cleared and the tools to do it being manufactured in numbers.

Studies of ancestral DNA strongly suggest that farming and all of its techniques was brought by immigrants, most of them men. The most common Y chromosome marker in Britain, found from Shetland to Cornwall, is R1b, and its wide distribution and overwhelming numbers showed how successful the first farmers were. Our south-facing paddocks run down to what was the Hartwood Loch and because they were free-draining they were good places to plant the early, primitive cereals cultivated by the R1b men.

The settled peace and beauty of this place has filled the hearts and eyes of five hundred generations of hunters, gatherers and farmers, and sometimes in the early morning or the late evening, when the veil between worlds is thin, I feel them walk beside me.

28 January

Perhaps economic circumstances will force me to follow in the furrows ploughed by my ancestors but for the moment what I do is better described as small-scale husbandry. Using the wood that grew here, and having planted about three hundred trees since we arrived, I am husbanding that resource at least. And by beginning to grow garden crops in a small way, I am using the land to support us. Leaving aside any matters of principle or continuity, what I grow, the potatoes and tomatoes, taste a great deal better than anything that can be bought in a shop. I enjoy watching plants grow from seed and, in the future, I shall expand our raised beds and indoor planting.

29 January

Porridge changed the world. This is not an absurd Scottish nationalist fantasy but the conclusion of an attractive conjecture, something I have been thinking about as I work on the farm and try to bring life to its long past.

It is very probable that hunter-gatherer populations expanded extremely slowly, often only replacing generations, each couple having only two children. There were compelling reasons for this. Families found it difficult to move quickly and easily around the summer Wildwood with more than one or two children in tow. If one little one could walk, the other could be carried papoose-style, like Native Americans.

The second reason is more complex. The hunter-gatherer diet of meat, roots, fruits and berries is difficult for baby teeth to chew; some of the mothers in the tribes of the Amazon masticate their toddler's food to break it down before putting it into their hungry mouths. For many millennia, the principal source of protein was breast milk and it is likely that prehistoric babies and toddlers were suckled for much longer than they are now, perhaps up to four years. Women are generally infertile while nursing young ones and so the birth interval in hunter-gatherer society might have been long. Given that the fertile lives of women were brief, with most dying at about the age of twenty, it seems very probable that two or at most three babies (who survived) might be born in each generation. For aeons, our ancestors flitted through the Wildwood, barely rustling the leaves of five thousand autumns.

When primitive cereals were grown, it is likely that, as well as making flour for unleavened bread, the pounded ears were mixed with water or animal milk to make a version of porridge, a loose paste that could be fed to young children that was high in protein and did not require to be chewed.

This in turn meant that babies could be weaned more quickly and that even in their brief, fertile lives women could have more children. The population began to grow rapidly and this was, I believe, the most important effect of the introduction of farming. The south-facing slopes of our fields, catching most of the sun, would have been good places to grow the small yields of cereals needed for prehistoric porridge and bread.

The other vital skill brought by the first farmers was the ability to domesticate animals, principally sheep and cattle. Both produced milk and once again this was a valuable food resource, especially for children. More than ten thousand years ago, human beings lost the ability to digest milk after weaning, but the early farmers overcame this by a process of natural selection and lactose intolerance was much reduced. This gave rise to a largely stock-rearing economy. I believe that our ancestors grew few crops but raised many sheep, cattle and some goats. The landscape around us still favours pastoralism and, like most of my neighbours, we mainly grow grass.

The dependence of the first farmers on their animals shaped a different sort of year, one that revolved around their rhythms, and they lived by what might be called the cattle clock. When pregnant cows and ewes began to feed their young, it signalled the end of the winter, and when the ancient journeys of transhumance led beasts and men up the hill trails, summer was starting. To relieve pressure on the inbye fields, herds and flocks were driven up country to fresh pasture, where shepherds and their dogs lived out with them, mainly protecting young animals from wolves. They lived in temporary structures called shielings. Our little valley is so perfect for these ancient practices that my neighbour still moves his stock around in similar ways. From the fields around Hartwood Loch, flocks and herds were driven up to the western hills behind Hartwoodmyres. And up in the Deer Park there are two scoops cut out of the limestone which look as though they might have served as shelters or shielings. There is no other reason for them being there.

30 January

In the circle of firelight faces flicker, staring at the flames. Sheep pelts from the autumn kill are pulled tight against the chilly draughts and the cone of sticks and split logs crackles as the father pokes at them. Wool combed from the summer ewes has made yarn woven into warm wraps for the little ones. Outside, the winter winds whistle and the world shrinks. Stories are told: how last year's lambs sat on their mothers' backs and how the calves cavorted and broke down the pens, how a coracle capsized on the loch and all the eels escaped, how springtime would come, and always tales of the spirits of the woods and the water.

As the little ones become drowsy, hunkering down below the lingering smoke, the father piles the fire with slow-burning whole logs and the flames die down to embers. In the chill of the long nights, the mother wakes often and turns the logs so that the embers glow and the hut warms a little.

Each morning I light the smaller woodburner and fill it with split logs left in the basket from the day before. Once it is roaring, the dogs, full of their breakfast, lie down on the sheepskins around it. My mood lifts as the dancing flames burn yellow and the wood crackles. The most urgent chore is to go out to the log barn to fill up the baskets, and this morning I had to use the axe to chop logs smaller so that they burned faster. Behind a row of orange Scots pine, I found an abandoned mouse nest the size of a shoebox. Its intricately woven moss and horse hair must have been the work of many days, the woodland mice gathering it up and slipping through to the back of the log piles to their secret home. Given the breeding rate of mice, it had been home to many. When I took the split logs that had been stacked around it back to the house, Maidie became very excited. But when she found no signs of prey, only their pungent scent, she lay back down in front of the fire and dreamed of rabbits in the springtime.

31 January

It was so cold this morning that the metal field gates were sticking and had to be kicked to free the catches. A low, dense grey mist muffled the land and hid the tops of the old sycamores. I could hear roosting crows cackling, nagging and shifting, but they were invisible. Even though there was no snow, a deep frost had made all white, stiffening the grass and turning the hawthorn hedges ghostly, a graphic, filigree tangle. Out in the East Meadow, the Old Boys and the mares stood motionless, heads down, recovering from the chill of the bitter night. Only the red blinking light of the electric fencing relieved the monochrome landscape. All was still, silent, waiting.

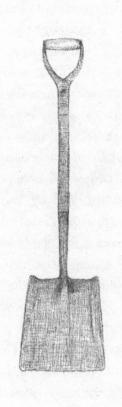

February

1 February

The land remembers the first farmers. Its contours carry memories of change and of continuity. Six thousand years ago, when the prehistoric peoples of our valley began to clear the tangle of the Wildwood, their great labour made them choose with care. Hacking down trees with the razor-sharp Langdale axes and burning back the bush and scrub was only worthwhile if the little fields were free-draining, sheltered and south-facing into the circuit of the sun. Our grass park below the Top Track and my neighbour's beyond it were good places to grow barley and oats. They now drain down to the stream known as the Common Burn, our southern boundary. When the first fields were cleared, the run-off would have gone into Hartwood Loch.

When the new geometry of fields and enclosures was created, the effort involved must have changed attitudes to the land. Rather than the customary rights to their ranges, probably only rarely defended by the small bands of hunter-gatherers in the sparsely populated landscape, the farmers developed a firmer sense of ownership. None of this is more than conjecture, but it is likely, given the sustained effort required to make the fields.

No one knows what language the new farmers spoke, or indeed that of the peoples they largely supplanted. My own belief is that they brought early versions of Celtic languages, what evolved into Gaelic and Welsh. Whispers of the names they gave their places may survive on the modern map. Long before Gaelic came, dialects of Old Welsh were spoken in Scotland, and not far from our valley are some echoes of a different linguistic past, a different way of seeing a wilder, arboreal world. Near the village of Maxton is Pirnie, and it simply means 'the wood'. Cognate is Primside in the foothills of the Cheviots, 'the settlement by the white tree', and across in Berwickshire is Printonan, 'the wood by the moss or the bog'. There is not a shred of evidence to support this, but I like to think something like Printonan was the first name for our farm.

2 February

For millennia we have looked upwards. Each morning the enduring drama of the sky – the source of light, warmth, storm and lightning – has made us all pause to lift our eyes, not only farmers but millions of commuters and city dwellers on their way to work, on the school run, walking a dog or putting out the rubbish. We famously use the weather as a greeting or a way of opening conversation, but what we are really talking about is the sky, the wide blue yonder.

The place to see it in our valley is the Howden Motte. Steep-sided, it is a promontory that sits at the western end, overseeing all, commanding very wide vistas, about a mile from our farmhouse. Not too far for my little dog, I decided to walk up there and test a new interpretation. Called a motte because it was thought to have been the site of a Norman motte-and-bailey castle of the eleventh or twelfth century, it might in fact be much older. From a visit made some time ago, I had come to believe that it was originally a Bronze Age hillfort, dug and enclosed around 1500 BC.

To reach the old track that shelves into the flanks of the promontory the motte sits on, Maidie and I had to ford the Hartwoodburn. Very unwilling to splash across, her undercarriage being low to the ground, the little terrier dug in her paws on the bank and I had to lift her across. Embarrassing.

Once on the winding track I could see that the site is tremendously impressive, with very precipitate slopes to the south and north, and a gentle ridge leading from the east and up out of our valley. To the west the land falls away sharply down to the burn and the sparkling, tumbling waterfall of Motte Linn.

I could see that the site had been clearly man-made or man-enhanced, earth mounded up to make it even more commanding. But immediately on passing through the western gateway I realised that this was no Norman motte. There is a perfectly preserved example in Hawick, only ten miles away, and I know it well. It resembles an upturned pudding bowl, very simple and effective in its construction, designed as a refuge of last resort, resistant to cavalry, so steep it is difficult to scale. But at Howden, clearly visible through the leafless trees and the winter dieback, was evidence of much more. This fort was multi-vallate, with at least two more ditches visible on the steeper sides and probably others that had been ploughed out in the fields to the west and east. And certainly from the gentle slope in the east there would have been little protection from a concerted charge of cavalry. The top of the site was also too large for a Norman motte. Hawick and others I have seen are much smaller and more restricted. This place had been built not a thousand years ago but three thousand.

Surrounded by a screen of hardwood trees that have sent deep roots down through the looser soil of the banks and ditches, the interior of the hillfort was flat and looked to be about an acre in area. A low bank around the edges seemed to me to have been a later adaptation of this wonderfully well-positioned site, and in 1302 there is a record of a 'fortalice' built near Selkirk whose description fits. Beyond an eastern gateway that may not have been original, there are the ruins of a rubble-built cottage, perhaps eighteenth or early nineteenth century. A strange place to build a house, with a water source a long way away at the foot of a steep slope, it seemed a bleak location on this windswept day. Maidie climbed and scrambled over the fallen stones, looking for rabbit holes.

Returning to the western gate and the long vistas to the hills of the Ettrick Forest, it struck me that this deserted, overgrown and tumbledown place had a powerful sense of drama. A freshening west wind blustered my face and I saw what I thought was rain sheeting in from the south. What made the motte atmospheric, a place of spirits, was not the promontory itself but what it showed all who climbed it. Here is where the first farmers came to look up at the vault of Heaven, the life-giving, sheltering, angry and unforgiving sky, and to wonder what celestial forces made its moods. I shake my fist in frustration at the winds, cursing out loud the storms and the damage they cause. Perhaps my wiser ancestors came up from the valley to this high place to pray for sun, warmth and life.

3 February

In winter, my gaze is often drawn to the motte. Perhaps it is an ancestral instinct. Through the stands and shelter belts of leafless hardwood trees, I can see what a focus it once was in our little valley, like a castle on a hill. It lies about a mile west of our farm. Three wide fields straddle the ridge leading up to the promontory and its prominence was hidden by agricultural improvers. On the margins of each field, trees were planted and, after two hundred years, most stand tall, their leaves lush from spring to autumn. Some are specimens, saplings planted for their looks as well as shelter. Each year, when the winter frosts have barely abated, the first hardwood whose buds unfurl is a Corstorphine sycamore, its vivid, lime-green leaves forming part of the screen that hides the motte from its valley.

Four thousand years ago the landscape looked very different.

After a period of advances and reverses documented by an analysis of ancient pollen, a continuous assault began on the Wildwood and the acreage of pasture expanded steadily. Evidence of wood-craft was dated to sometime around 2000 BC. With stone and flint, wood was a vital resource and trees were farmed by our ancestors. Coppicing, the cutting back of a central trunk to stimulate the regrowth of suckers around its base, was developed at that time. Preserved in anaerobic mud, the limbs of coppiced trees were used to make causeways over boggy ground or jetties into lochs like Hartwood. Our valley is now shaped and dominated by its woods and trees, but in 2000 BC it would have been much more open. And the motte would have been visible from all parts.

Four thousand summers ago, on the day of the solstice, a procession may have made its stately way up the central, spinal ridge of our valley, perhaps singing as they went. When they reached the promontory, they would have looked out over a landscape basking in the warmth of the longest day. On the hills above Hartwoodmyres their cows and calves, ewes and lambs would have grazed the sweet young grass, growing sleek on its succulent sugars, and below them green crops grew in small enclosures close to their huts.

The cultivation of land had led inevitably to a sense of owner-ship and in turn social hierarchies were established. The people of the valley would have processed up the ridge on command. The Lord of Hartwood, or at least a leader of some kind, prob-ably a man who also assumed a priestly role, would perhaps have held a ceremony to begin the work of building his capital place, a fort and also a refuge for his people. It was in the time between planting and harvesting, when stock had been driven upcountry and the inbye fields were empty and recovering. It was a time when work could begin.

It is likely that several teams began to dig simultaneously around a perimeter marked out in a ceremony of some sort. With antler picks and baskets, they would have excavated ditches and piled the upcast on the summit of the bare promontory. Once the teams had linked their lengthening ditches into a completed oval shape, the work of building the palisade would have begun. A timber frame was rammed into the ground, essentially a cage into which any stones and the upcast would be piled. Once the rampart was complete, the builders would have brought cut timber to form the palisade or stockade itself. Larger tree trunks were made pointed at one end and rammed into the excavated earth before being braced and the gaps filled with shorter stakes cut from coppiced trees.

The western gateway was the weak point and the rampart was constructed so that an assault was funnelled through a narrowing entranceway before reaching the wooden gate. This small hillfort could have been completed in a single summer by work gangs of twenty or thirty people. The motte was the first building in our valley of which any trace remains and its creation gives a sense of how many people lived on its small farms. Perhaps there were sixty in total, taking account of children and those too old or infirm to work. Now, only half that number live in the old lordship of the motte.

No archaeology has ever been done on the promontory, only some surveys, but all of the conjecture above is constructed on solid foundations. In 1931 an unfinished hillfort in Hampshire was excavated so expertly that the way in which it was built and how long it took could be accurately recreated.

Once all was complete at the motte, no doubt another ceremony would have taken place at the head of our little valley. At the summer solstice, Maidie and I shall go back and make an offering to the shades of our shared past.

4 February

High places lift us up, bring us closer to the gods. We climb to leave behind the ruck of the world below us so that we can gaze on the majesty of Creation. When looming storm clouds, heavy with rain, collide, and thunder rumbles and lightning crackles, it is not difficult to imagine divine hands directing the heavens. The Greeks and Romans, rationalists both, believed that Zeus or Jupiter threw down thunderbolts, and Thunor, the Anglo-Saxon god of storms, gave his name to thunder. When he was angry, the god smote his anvil hard with a huge hammer and thunder boomed across the sky and lightning crackled.

Long before Zeus roared, it seems certain that our prehistoric ancestors looked upwards to find their gods. None of their names survive, their rites are mysterious, but the spectacular sites of their worship can still be seen in the landscape. Misleadingly called hillforts, hundreds were dug in the high places of early Britain. Many are too large to be defensible. Visible from the Deer Park, Eildon Hill North's banks and ditches describe a circuit of a mile and have five gateways. It would have needed a garrison of thousands to man the ramparts. Other hillforts are overlooked by higher ridges from which missiles could rain down. Although some were fortified in times of war, especially during the Roman invasion after AD 43, the principal role of hillforts was religious. They were sky-temples. In a modest way, the motte was almost certainly a focus of worship for the people of the valley.

Modern thinking generally divides church and state, but in the past no such distinctions were made. Roman emperors were routinely deified and pharaohs and priest-kings ruled over some early societies. It seems likely that his people believed that the Lord of Hartwood knew the minds of the gods well, and at the motte I think that ceremonies took place at the turning points of the year: the first fruits of early spring when the ewes let down their milk, the ancient journey of transhumance at the beginning of summer, the harvest and the cull of animals before winter. It seems that fires blazed on the motte. Names remember those nights when flames rose in the darkness. Tinto Hill near the Clyde translates as the Fire Hill and Carntyne as the Fire Cairn, both places of ancient ceremony.

Roman consuls and generals rarely acted before their priests had interpreted the auguries in the sky. These often involved observing the flights of birds: their direction, number and when they took place. This is likely to have been another ancient practice and one best undertaken from the vantage point of high places. Interpretation was, of course, everything, but canny priests knew the migration patterns of different birds, and also had more rational, political means of judging the consequences of actions. The birds could have meant what the priests or the generals wanted them to mean.

5 February

Hunting in our valley has never ceased. Even after farming arrived six thousand years ago, it continued. It had to, for in the hungry month of February, when winter stores were running low, birds could be killed with an arrow – a cleaner death than being riddled with shotgun pellets. It is a continuity of sorts – the pop of cartridges regularly punctuates our winter in the valley. The reality is that almost all hunting is now recreational and not born of necessity. I hate to see animals killed for sport, and although I am far from being a vegetarian, shot pheasants and partridges are not something I would eat. My squeamishness is reinforced by the danger of biting on a lead pellet and breaking a tooth.

6 February

At minus two overnight and the gauge climbing, it felt almost balmy as Maidie and I ambled up the track, able to lift our heads from watching out for ice patches. And when the first gunshot cracked in the still morning air, we both jumped and I instinctively cowered. But when I realised it was the sharp report of a rifle and not the crump of hunters' shotguns, I understood what was happening. At Windy Gates a silver pick-up was parked and across the back seats there was an empty gun slip and a box of ammunition; high-velocity bullets, not cartridges. A marksman had come to shoot the young deer. My neighbour had told me a few days before Christmas that there were too many roe deer in the woods and that they would have to be culled. When it came to planting time, the hungry animals would make a real mess in a barley field full of succulent shoots.

But where was the marksman? The Old Boys and the mares were spooked but not panicked by the gunfire. I wondered if he had hidden himself in the Young Wood or the shorter trees of the New Wood beyond the grass park. I doubled back to Windy Gates and to my amazement saw the roe deer stag, showing himself in full view in the open field. He was sniffing something lying on the ground. It was the body of one of his children. When the marksman emerged from the fringes of the New Wood, the stag froze, and then raced downhill towards the Hartwoodburn and the safety of the trees.

Dressed in highly camouflaged kit, carrying a tripod and a rifle with a telescopic sight, the shooter began walking towards Windy Gates. At first he was very defensive, 'I don't want any conflict' and 'I stopped when I saw you.' When I explained that I understood that the deer needed to be culled, he relaxed a little and explained to me that he'd gone behind the New Wood and through the upper part of the East Meadow so that he could position himself downwind.

Anxious to get on, he opened the field gate and drove to where the stag had been standing over the dead youngster and another carcass. There had been only three shots and I watched him drag by the back legs the two young deer he had killed over towards the pick-up, next to the fence at the New Wood. He bent over each carcass. I realised that he was gralloching them, gutting each one and throwing the innards, the liver, kidneys and lights into the dieback. There would be a feast for the foxes, the crows and any buzzards that could elbow their way in. The scent of the guts, the stench of death, would be in the air all morning.

When the marksman drove back to Windy Gates, I waited to ask him if and when he would be back. More relaxed, he gave me his mobile number and told me his name. He turned out to be far from a cold-hearted professional hunter. Liking deer, he was sad to have to kill them and, as I had seen earlier, he was sure I was going to give him a hard time. Perhaps others have. He told me he spent four months a year in Norway shearing sheep and in the Borders made a living as a stalker at the many pheasant shoots, as well as shooting deer. When he said goodbye and gripped the steering wheel, I noticed that he had dried blood around his fingernails and between his fingers.

7 February

When morning mist muffles the land, it seems to descend into the dark deeps of the world. Like reefs or wrecks, stands of trees loom out of the grey silence. Then, when the breeze shifts, in moments they disappear. Unseen, the sun climbs out of the east and, after a time, there is a patch of blue overhead, like the surface of the ocean seen by a diver swimming up to the light. In this waking dreamland, the breeze shifts once more and in the half-world I shiver at the wraiths swirling in the folds of the mist. Perhaps I felt a tap on my shoulder.

9 February

Last night history flooded back across millennia. After a long and steady downpour the Tile Field was drowning once more. Several ponds had formed on the lowest levels and over to the west a wide area was inundated. Five winters ago, we had a long period of intermittent rain and the ponding was so widespread that it was possible to see where the ancient margins of the loch were. I watched a flock of gulls feasting on the drowned worms.

10 February

The hills remember the past and the lowlands forget it. On the high ridges above Hartwoodmyres and Brownmoor in the south, the Ordnance Survey marks settlements, enclosures and forts, faint folds in the ground where banks and ditches were once dug by our prehistoric ancestors. They survive because they fall on the far side of an ancient frontier. For millennia, herdsmen have left the ground undisturbed, their beasts grazing, ewes making sheep-lawns amongst the gorse and the tough marsh grass, cows devouring even the roughest of pasture.

Our farm lies astride that frontier, the divide between herdsmen and ploughmen. When the brilliant Berwickshire blacksmith James Small invented the modern swing plough in the late eighteenth century, farmers could delve deeper, drain their fields and destroy almost all trace of ancient settlements. All that remained were the flints and other objects that the Mason brothers and Walter Elliot picked off the crests of the furrows.

Long before Small's earth-breaking invention, this faultline in the landscape was evident and marked on maps. When the Lord of the Motte looked out from his ramparts over the hills in the west, he gazed at the wild land, the territory of the Hunters. A map made in the second century AD described a landscape established long before when it plotted a kindred known as the Selgovae. It is a Latinised name derived from the Celtic root-word *seilg*, which means 'to hunt'. The settlements beyond Hartwoodmyres were the farmsteads of herdsmen who also hunted deer, wild boar and the giant feral cattle known as the aurochs, as well as packs of wolves and the solitary lynx that might prey on their flocks and herds.

All except the deer have gone, but the invisible frontier remains. Farming in the hills is still a much harsher life than ploughing the fertile fields of the Tweed Basin, but the old prejudices seem to have faded. When I was growing up in Kelso, time seemed to dance to a different rhythm for the shepherds who came into town for a Saturday night at the pub. A famous story tells of one who was seen standing, slightly unsteadily, at the bus stop at 9.30 p.m. for the 10 p.m. bus back up the valley. When told it would not come for half an hour, the old shepherd replied, 'Aye, son, it won't take me long to wait half an hour.'

11 February

The noises of the night were echoing around the valley. Out in the early dark with Maidie, we heard a hoolet call in the Hare Wood. 'Owl' in English, hoolet seems a more expressive name. My grannie used to make us laugh because she could hoot like a hoolet, and a rough transliteration might be hoolie-gooloo-oo-oo. Our hoolet's call was answered from somewhere on the northern ridge behind us, perhaps the wood around the Haining Loch. They usually call to mark their hunting territory. I am not certain, but they might have been tawny owls.

In the moonless darkness, these calls seemed timeless, something that had been heard across millennia by our ancient predecessors. It struck me that the Scots lexicon for bird names might also be ancient; they are so different from the English versions and some of them are onomatopoeic. Whaups are curlews, yorlins are yellowhammers, bubbly jocks are turkeys – all names derived from their calls. Others seem descriptive: hoodie for a carrion crow, corbie for a raven, laverock for a lark. And some are just very different, like gled for a buzzard.

Breasting a horizon clear of clouds, the sun rose quickly and the hoolets fell silent, settling down for their daytime roost. On a very cold morning, the warmth was welcome. Wind, rain and cold killed our ancestors, seeping into their bones. It can be no surprise that across the Earth the sun was worshipped.

The weather governed lives until the Industrial Revolution and the coming of indoor work for most people, but it will surely govern us once more as climate change accelerates. A headline in today's paper was profoundly alarming. The insect population is being devastated by pesticides and other factors so extremely that it is declining by 2.5 per cent a year. If nothing is done, it will soon be too late to avoid a catastrophic descent towards what scientists are calling a sixth mass extinction event. Except it will be an extermination. And there is little or no political leadership that even recognises what is going on, never mind having the motivation to do something before it is too late.

12 February

With only sixteen days of February left to endure, I decided to risk going out with Maidie without a jacket, me not her. It was dry and overcast but absolutely still, and with three layers I was warm enough. Another first, another sign that the weather is improving and that there will be light at the end of the dark tunnel of winter.

We heard them before they appeared. Suddenly a flight of eight geese honked above us, having only just cleared the treetops of the New Wood. Flying in a wide circle, they seemed to be searching for something, and ten minutes later we saw what it was. Very high in the morning sky, a spectacular double chevron of hundreds of geese was moving almost due north over the Ettrick and towards the Lammermuirs. Despite the distance, we could hear them honking and Maidie sat down to watch the progress of an epic journey. The chevrons seemed fluid, constantly changing formation, joining up and then drifting apart. Perhaps the honking was important for communication, to keep the huge flock together as they moved fast. This must be one of the most beautiful, most breathtaking sights in the natural world. It might be another sign that winter is loosening its grip if these birds are moving north again.

13 February

Almost two millennia ago news blew north on the wind. In the summer of the year AD 43 the people of the valley got wind of momentous events in the south. The most powerful man in the world had sent his armies to conquer the holy island of Britain. And when they had defeated the kings of the southern kindreds, the Emperor of Rome had ridden in triumph through their capital place, it was said, on a huge creature the like of which had never been seen before. It was called an elephant.

Like the animals of the Wildwood who sniffed the air for the scent of predators, we still say we 'get wind' of things to come, often rumours, strange stories, and there can be no reasonable doubt that the news of the Roman invasion crackled like wildfire up and down the length of Britain. There can be no doubt that the farmers of our little valley knew that the Empire had crossed the sea and moved north. 'Have you heard?' ricocheted around the western hills. It was an event that had been planned for months and was anticipated in much more remote parts of Britain.

In AD 51, the Emperor Claudius had a triumphal arch raised in Rome that listed eleven British kings who came to Colchester to submit to him. One had come a very long way, more than six hundred miles to bow before the imperial throne. In advance of the invasion, the King of Orkney had entertained Roman diplomats and agreed to allow the islands to become a nominal part of the Empire. It was a vivid gesture, a demonstration that allowed a politically useful boast that Claudius's power could reach across the ocean to overcome the peoples who lived at the ends of the Earth.

In the fields around our farm, the Mason brothers had found evidence of wide-ranging networks of trade, and merchants always add news to the bargain. In the paddock near the farmhouse Walter and Bruce picked up two fragments of jet, a black, lustrous stone that was mined mostly at Whitby on the Yorkshire coast. Some of the flints Walter Elliot gave me came from Ireland and the Lake District, and, just as surely, stories of startling events, of gigantic creatures and vast armies of uniformed legionaries came up from Colchester and the south-east. In the same way that we are, our ancestors were especially curious about breaking news, and each traveller would have been interrogated for the latest developments.

In the first millennium BC, politics in Britain had shifted focus. Warlords who commanded bands of warriors began to carve out small kingdoms and only a few miles to the east, on the northernmost of the three Eildon Hills, a huge building project had begun. Around the summit a vast rampart was dug, more than a mile in length, and on a shelved plateau on the southern flank of the hill three hundred roundhouses were raised. This vast hillfort was the capital place of powerful kings and it sits astride the ancient frontier between the shepherd-hunters of the Selgovae and the ploughmen-farmers known as the Votadini. Which kindred controlled the great hillfort is a matter of continuing conjecture.

A cold wind blew this morning, bending the young trees by the Bottom Track, and I thought about the long past as I pulled up my collar. Just as we do when trouble arises, the people of the valley would have thought of the Roman invasion as a threat, but comfortingly remote. We see our farm as a refuge from the storms of the world, and we hope we keep the gathering chaos at arm's length. Two thousand years ago the tramp of the legions, drums beating, harness jingling, was a distant rumble of breaking thunder, but it did not remain so for long. Perhaps that is a lesson of history we should learn.

15 February

On a spring morning, at the foot of the Long Track where it crosses the Hartwoodburn, the army of the Empire marched. Between six and seven thousand men – Roman legionaries and auxiliaries from the northern provinces of Batavia and Tungria, all in full armour, their shields slung, the eagle standards glittering in the sun – were making their way west into the hills to build a fortress. Flanked by detachments of mounted scouts on the northern and southern ridges of our little valley, the soldiers were probably led by Gnaeus Julius Agricola, their general and the governor of Britannia.

Almost forty years after Claudius came to Colchester, the Emperor Vespasian had ordered the conquest of Caledonia. With detachments from the II, the IX and the XX Legions, Agricola advanced north in a pincer movement to surround the hostile Selgovae. One invasion force moved up the line of the modern M74, and the other up the line of the A68. In the lee of the vast native hillfort of Eildon Hill North, the legionaries built a large fort and depot. Once that base had been established, their general decided to strike into the heart of Selgovan territory, and on that spring morning nineteen centuries ago the jingle of harness, the creak of ox-carts, the thud of hobnailed sandals and the shouts of the optios to keep their centuries in formation rent the air of our valley. Rome had come and history had begun. Hard, recorded, reliable facts chased the suppositions and uncertainties of prehistory into the shadows.

I was out with Maidie on a crystal morning of brilliant sun, and I decided that we should go and walk in the footsteps of the legions and look at what remained of their presence in the western hills. In a slanting winter light, I followed the line of an old drystane dyke up to a plateau topped by a rectangular plantation of sitka spruce. This was the site of Oakwood Fort, about three miles west of our farm. Only discovered in 1949 by a sharp-eyed surveyor looking at aerial photographs, it is a mysterious, unsuspected place. I had hoped that the low sun would show up the shadows of banks and ditches amongst the tussocky marsh grass. But there was virtually nothing to be seen. On the east and south sides all I could make of the place was a plateau with gently sloping sides and on the north and west what might have been ditching. But, to my eye, they looked more like natural features.

When in 1951 and 1952 archaeologists surveyed the area and lifted the turf, they uncovered an extraordinarily rich record. To the north of the intended site of the fort was a temporary camp of thirty-three acres built to protect Agricola's legions and their auxiliaries while they worked on the defences of the fortress. In a more or less square design, they raised a rampart of between eighteen and twenty-three feet in width by piling up turf, using the sods like large bricks. On top was a palisade of rammed stakes and at each of the four gateways were two high towers or fighting platforms on either side of double portals, one gate for entering and the other for leaving. The oak stumps of these towers were found by the archaeologists to be still in situ. Each gate and its towers was set back thirty feet from the line of the rampart to create a funnelled entrance way, just like at the fort on the motte. It made attackers vulnerable, forced to take fire from two sides.

Oakwood's three and a half acres housed five hundred soldiers, a legionary cohort as well as a mounted detachment. It was placed on this site because it is possible to see the summit of Eildon Hill North through a gap in the eastern hills. There the garrison at the large fort known as Trimontium built a signal station and messages could be exchanged in moments.

As the Selgovan warriors watched from the high ridges, Rome was planting its standard in the heart of their territory. Seen from long distances up the Ettrick Valley to the west, this fortress was a mighty military symbol, garrisoned by professional, hardened soldiers protected by cavalry. It was an early example of a glen-blocker fort, the sort built to contain the mountain kindreds of the Highlands when Agricola's army marched further north.

Contact with native farmers and herdsmen must have been constant. With more than five hundred mouths to feed, and with horses to graze and find winter forage for, the fort's quartermaster bargained with local food producers. There are five native enclosures, ramparts that were probably palisaded, within a mile of the garrison and in two of them Roman coins have been found. Fortifications as close as that would only have been tolerated if they were useful and friendly. Oakwood was occupied for only twenty-four years. In 105 the Emperor Trajan recalled legions from Britannia to fight a flaring war on the Danube frontier and the Romans abandoned their conquest of Caledonia. Archaeologists found evidence that the wooden towers had been burned. Perhaps the kings of the Selgovae made a bonfire of their humiliation.

I know that the legions of the Empire marched past our farm because in a brilliant investigation of the ground over considerable distances archaeologists found the road they built between Oakwood and Trimontium, the army depot below the Eildon Hills. There is evidence that it was used much later as a boundary between farms and in places the road-mound is very clear, sometimes as much as five metres in width. The soldiers at Oakwood spent almost all of their time not fighting. Road-building kept them busy and out of trouble.

16 February

This is the time of the year the native kindreds called Imbolc, the feast that celebrated first fruits, when ewes let down their milk before lambing. It was a welcome signpost in an annual cycle of change and renewal. This morning I saw straw bedding being taken into the lambing shed up at Brownmoor. In two or three weeks' time, depending on the weather, the first of the little ones should be staggering around their pens, blinking under the heat lamps, bleating for warm mother's milk. Perhaps my granddaughter will be old enough to see the wee lambs.

17 February

I love the early mornings, the hour when the darkness slowly dissolves, the day begins and the mind clears. An open sky promised sun, and Maidie and I climbed up to the ridge where the Top Wood once stood. Because she is little more than fourteen inches off the ground, the terrier likes to jump up out of the tall grass onto a tree stump and see a little more of the world. The ridge commands wide views of the whole valley, and long vistas to the west and east. While we waited for the rays of yellow warmth to rise over Greenhill Heights, seven swans flew over us, no more than forty or fifty feet up, so close I could see the orange of their beaks and the black markings above them. The great birds were honking, as though encouraging each other to fly faster, stay together, and I wondered at their urgency. Where were they going in such a hurry? And why?

As the sun bathed the land, rising very quickly, and the dieback glowed gold, the seven swans flew west. I could see the plantation of sitka spruce at Oakwood Fort peeping over the far horizon and wondered what the sentries would have made of these majestic birds. There were seven, a magic number, and perhaps they muttered to each other that this was an omen, good or bad.

First light in the fortress would have followed a strict military routine. In the principia, the headquarters building that stood where the scruffy sitka now grow, the fort's centurions would meet for morning report with the prefect. They submitted lists of men available for duty, and those who were absent or sick were recorded. The password of the day was agreed, the sentries around the garrison's standards were named and orders were given. At Oakwood, patrolling and the gathering of intelligence in hostile territory will have been a prime concern and reports will have gone regularly to the regional headquarters at Trimontium. Work rosters were drawn up. The fabric of the fort needed constant maintenance, materials had to be sourced and roads built, kept passable, especially in the winter. Mounted messengers would have moved between Oakwood and Trimontium, and heavily escorted packhorses, ox-carts and mules laden with supplies would have creaked along the new road at the foot of the Long Track, half a mile from our farmhouse. When I cross it, I feel an intersection of history.

Standing with Maidie on the ridge of the Top Wood, and knowing how echoic our valley is on still mornings like this, I am certain we would have heard the tubicen, the trumpeter, sound orders as legionaries and auxiliaries left the fort on patrol or drilled on the parade ground by its walls. Highly organised, well trained and ruthless, the Roman garrison dominated the valley and the hill country around it. The contrast between these uniformed soldiers, with their red cloaks, shining armour, plumed helmets and disciplined marching columns, and the native home-spun of the warriors of the Selgovan kings could not have been more stark.

For twenty years our valley was drawn into the Empire, a highly connected wide world that stretched south to Africa and east to the deserts of Persia. The cohort in the fort may have been Spaniards from the IX Legio Hispana, or Gauls from II Legio Augusta, or perhaps Italians from the XX Legio Valeria Victrix. When Oakwood was built, Rome was in its pomp, its Emperors masters of the known world. But it would be a mistake to confuse the Empire with civilisation.

Because the Romans left written records, were spectacularly successful in warfare, created an Empire that lasted until 1453, when Constantinople fell, were ingenious engineers and, most important of all, were an urban culture, historians give them a disproportionately prominent role in our history. Indeed, many histories of Scotland and Britain begin with the Romans, devoting only a few introductory pages to prehistory, the story of our ancestors, those who peopled our landscape for eight or nine millennia before the Roman armies brought slaughter and destruction. It seems to be forgotten that the soldiers at Oakwood and across Britannia were colonists, exploiters and oppressors. Just because the early history of Britain is difficult to piece together does not mean it should be ignored.

18 February

It rained so heavily this morning that even taking the dogs out to pee earned me a soaking and a first change of clothes. So much of our clayish mud stuck to my boots that I had to stand even longer in the downpour to scrape it off. The Anglo-Saxons called February *Solmonath*, Mud-Month. It is listed in a text that fascinates me. In addition to much else, and his magisterial *Ecclesiastical History of the English People*, Bede of Jarrow wrote a treatise called *De Temporum Ratione* (*On the Reckoning of Time*) in the early eighth century. It was enormously influential and shaped the way our culture sees the passage of time.

Because he was the first genuinely scrupulous and generally accurate native historian in British historiography, Bede took great trouble with dates in his *Ecclesiastical History*. To make the sequence of events as clear as possible, he adopted the AD system of dating we use today. It was invented by Dionysius Exiguus (Little Denis), a monk who died in AD 544 at Tomis on the Black Sea coast. He worked out that AD 1 was the year when Christ was both conceived and born. There is no evidence that he was correct, and some that he got it wrong. According to the gospel writers, Christ may have been born in the last year of the reign of Herod the Great – that is, 4 BC. Or in the year of the first

Roman census of Judaea, which took place from AD 6 to AD 7. And so we may all be living in the wrong year.

Bede had other reasons for writing *De Temporum Ratione*. The early British church had been riven with dispute over the dating of its principal festival of Easter and Bede used the new system to work out a table of dates for Easter up to AD 1063. He also wanted to sort out a chronology of world history up to the reign of his contemporary, Leo the Isaurian, Emperor of Rome in the East at Constantinople. Although AD was a concept created by someone else, it was Bede's adoption of it that led to its use in Europe, particularly at the court of Charlemagne, and its ultimate ratification in AD 1048 by Pope Leo IX.

What became known as the calendar originally had nothing to do with dates. It comes from *kalendarium*, Latin for an account book, more precisely a moneylender's account book. It was specifically applied to the first day of the month, the *kalendea*, the date on which bills had to be paid and debts settled. This habit of monthly accounting is ancient and persistent. It gave rise to thirty days' terms for invoices and is still widely followed in modern business transactions.

All of these systems and nomenclature developed a sense of the march of time, that it was linear, a progression. And it also fed the notion that history really began with the Romans. The system of BC dating, counting backwards through the millennia Before Christ, only became current much later, in the seventeenth century. It was as though nothing much mattered before AD, and anything that did was only a prelude.

Paradoxically, religious belief offered another, different way of understanding the passage of time. No specific references to a Day of Judgement exist in the Old Testament until its twenty-seventh book, the Book of Daniel. This was written comparatively late, in the second century BC and it described how the dead would be resurrected on that fateful day. The most comprehensive account, as with so much that is now accepted as popular doctrine, is found in an apocryphal text, the Second Book of Esdras. Before the day itself, there would be a temporary messianic kingdom on Earth, then a week of primeval silence, and only then would the dead rise and be called to account. This event marked the end of an age and the beginning of a new mortal era. Such thinking challenged the linear model – and the history of the world, of humankind, was seen as cyclical. The term Middle Ages originally signified the middle age between Christ's first coming and his second.

Now, with the decline of Christian belief, the linear model dominates and is intertwined with ideas of progress, improvement and greater understanding. It is an attitude that might prove fatal for our planet as the evidence mounts that we are destroying it much faster than scientists believed possible.

20 February

I believe I was born with no distance in me, no detachment. I can quickly get close to my work and many of my enthusiasms have been lifelong. I have been too loyal to friends and sometimes been badly let down. But most important to me is family. In a moment of uncharacteristic harshness, my grannie once said, 'All that matters is family, the rest are strangers.' Bina was wrong, I think, but the more I discover about the mysteries of her early life, the better I understand her outburst.

One side of my family was busy, densely populated, vibrant. Raised in the textile town of Hawick, my mum was one of seven sisters and a solitary brother. Consequently I had dozens of cousins, and after dad had bought and done up an old banger we saw our Hawick family often. 'Cruising at forty,' he used to say with pride, as we drove up the Teviot Valley in the black Morris Series E. I remember going on holiday to Hawick, sleeping in a box bed in Auntie Jean's ground-floor flat in Gladstone Street. We all went to the pictures – on a weekday! It was the era when films were shown on a continuous loop and it gave rise to the phrase 'This is where we came in.'

My dad's side of the family was a blank. Beyond and beside Bina, there was no one, no gaggle of aunts or uncles, no cousins, first or even far removed. The silence was only broken after my dad died on a snowy night in February 1986. Later, Mum told me that he had been an illegitimate child, much more of a stigma in 1916, especially in a small town. But very slowly, over time, the silence gave way to whispers of the past and recollections of relationships.

Because that was a part of her life she told me something of, I knew that Bina was born at Cliftonhill in 1890, but it was a long time before I discovered that she herself had been the illegitimate daughter of Annie. No one then was alive who could tell me who my great-grandfather was. Two of Bina's aunts were also unmarried and had no children that we knew of, and so the Moffat line, from my great-great-grandfather William, came down only through my grannie to my dad.

When Bina died in 1971, I was away on a first long trip abroad, almost two months spent in Turkey, Greece and Italy with school and university friends. So as not to worry her, I had told my gran I was going to Torquay. When I came back, she wasn't there. Bina had died without me, and I'd had no moment to say goodbye and tell her how much I loved her. There was not even a head-stone I could visit. I suspect my mum and dad could not afford to pay for one.

That has always saddened me, and so yesterday I drove down to Kelso to see if I could at least find the lair where she is buried. It felt like time to do that. Having found no record online, I knocked on the door of Robertson's Memorials, a monumental sculpture business handily placed next to the cemetery. The lady was very helpful and gave me a number to call at Scottish Borders Council. Also very helpful and thoughtful, another lady told me that Bina had been buried in section G1 near the entrance gates of the cemetery. That was a flickering memory from the autumn of 1971 when I came back from Torquay. On my return to Robertson's Memorials the lady showed me a diagram so that I could see approximately where my grannie's grave was and also gave me an email address for the council. It was much better than nothing.

On my way back through the cemetery to my car, I stopped by my mum and dad's grave. I go every winter to see them and the tears always come, sadness at the loss of them and the passage of all that time. I know that they lived good and decent lives, but neither ended well. Their last years were hard, and that is what makes me weep. My dad was felled by two strokes in quick succession when he was fifty-eight and he changed completely from a powerful, assertive man to a limping invalid with a withered arm. After he died at the age of seventy, my mum seemed to lose motivation and direction. I remember her staring into the fire, silent, constantly smoking. Perhaps she thought she had nothing to live for. She was unhappy at the end of her life.

Looking at other headstones and recognising some of the names, I was stopped in my tracks by a headstone that remembered Andrew Hogarth. At the bottom, the name of his daughter had been added and she died in 2005. She was Robina Moffat Greig. That winded me. Robina Moffat was my gran's full name. In a small town, a mere coincidence seemed highly unlikely, especially with two relatively uncommon names. Who was Robina Greig and what was her link to my grannie? Was there a blood connection?

Once home, I emailed Scottish Borders Council and the lady came straight back with more information than I had bargained for. Not only did she pinpoint exactly where Bina was buried, she also added that Annie, my great-grandmother, and Isabella, my great-great-aunt, had been interred in the same lair in 1936 and 1928 respectively.

My sisters and I want to commission a headstone to remember these women who made us. But before we ask the mason to carve the names, another mystery needs to be solved. On a headstone in Ednam Kirkyard, about four hundred yards from Cliftonhill Farm, William Moffat is buried with his wife, Margaret Jaffrey, and their daughter, Mary – and, it says, their 'eldest daughter, Isabella, who died in 1931, aged 78'. Few people are buried twice. Who were these two Isabellas? No age is given for the Kelso burial, except that the records note she was the first to be buried in that lair. Having found my gran's grave and cleared up one mystery, two more sprang to the surface.

21 February

The past never remains in the past. Bina's memories, phrases, stories and the occasional unexplained, stray reference are slender threads in the darkness. Once, when we drove past a cottage in the village of Birgham, near Kelso, she remarked it was cousin Bella's house, but did not respond to immediate questions about her, nor did she want to stop and knock on her door. However distant a relative she might be, she was still a relative, the only one my gran ever mentioned. Alive in the 1950s, she cannot have been either of the Isabellas buried at Kelso and Ednam, but perhaps she was named after one of these ghosts. The void remained, and no candles ever flickered again in that dimmest and most distant of pasts.

In the morning mist, I searched the grass park by the Long Track for the grey shapes of grazing deer, hungry as they must have been even in this mild February. But since the marksman shot the two young ones I have not seen any show. Walking back to the farmhouse with visibility down to fifty yards, Maidie kept stopping to look behind her, back down the Long Track. But no one was following us.

25 February

A morning of contrasting senses. The air is filled with the sickly, sweet scent of silage drifting across from the cow byres at the farm and the sky is suffused with a gentle sunlight veiled by mist over Greenhill Heights. Yesterday evening I saw a roe deer hind, her white rump bobbing through the marsh grass in the Tile Field. Her camouflage is so perfect that it was only when she moved I saw her. The weather is very mild and grass is growing, tempting the pregnant does out of the cover of the woods.

27 February

The light is racing back. A pale dawn was quickly brightening and, for the first time this year, I could take the dogs out without a flashlight. After breakfast, Maidie and I climbed up to the vantage ridge to look at Creation. Winter sun changes the colours of Scotland as it warms the frosted fields from grey to green and lights the heather colours on Newark Hill. It will be another warm day, with temperatures up to sixty degrees Fahrenheit on the old scale, the one that means something to me. Human beings will bask again in the heat of a summer's day. But those creatures that hibernate will be stirring; hedgehogs might emerge too early and be unable to find the food they desperately need after the long months of winter starvation.

28 February

A dense and persistent grey mist has enveloped the land. It telescopes time and distance as the track behind disappears and shapes loom out ahead of us. Maidie stops often, her ears pricked, her head turning this way and that. The world seems both to shrink and expand. Because the sights of the twenty-first century are hidden, but the sounds are audible, the everyday takes on a different quality. The clang of the tractor down at the cow byres becomes an echo across eighteen centuries. Beyond the small pool of the visible, the clangour of ancient, half-forgotten battles rings out.

After the burning of the forts at Oakwood and Trimontium in AD 105, the warbands of the Selgovan kings rode east and south to raid Roman outposts. To contain these destructive incursions, army commanders at the legionary fortress at York concentrated large, thousand-strong units known as milliary cohorts in the west, around Carlisle. These were rapid reaction forces, a mix of cavalry and infantry, but they made little impression. By AD 115 to AD 120, it was clear that the warriors of the hill kindreds of the Selgovae, the Novantae in the east and the Brigantes in the south, were evading the cohorts, probably through their intimate knowledge of the hill trails, hidden valleys and the paths through the mosses and sykes of the upcountry. And when mist fell the Roman cohorts will have been at their most vulnerable.

By 119, a radical solution had been decided on. The commander of the fort near Carlisle ordered the construction of a wall of turf. It ran from Bowness on Solway to Willowford in the Irthing Valley, not far from Haltwhistle. Three forward forts were built – at Birrens in Annandale, at Netherby, north of Carlisle, and at Bewcastle in the eastern hills.

But even that was not enough. In 122 the hill peoples heard the booming thud of distant drums. The Emperor of Rome had come. Hadrian ordered a radical and permanent solution to the troubles of the northern frontier. From the estuary of the River Tyne in the east, a stone wall would be built across the waist of Britain to link with the turf wall at Willowford 'to separate the barbarians from the Romans'. It was the beginning of the idea of a distinction between the wild and savage north and the softer, more civilised south, between Scotland and England. The horse-riding warriors of the hill kindreds are the ancestors of the bands of Border Reivers who terrorised the countryside fourteen centuries later and who passed on their more peaceful traditions to those who ride the bounds of the common land around the towns of the Tweed Basin in the twenty-first century.

Hadrian's Wall was a vast, sprawling project that could not have failed to touch the lives of the people of our little valley. Its sheer scale sucked in resources from a very wide area. More than 3.7 million tonnes of stone were used, and for every ten men who worked on the wall itself, another ninety scoured the countryside for food and materials. To the north, the country of the barbarians excluded from the Empire, this process will not have been a peaceful negotiation, as Roman forage parties drove away their herds, appropriated their draught animals and emptied their granaries. Over five years of construction, thirty thousand carts and drivers were used, as well as six thousand oxen and fourteen thousand mules.

When the great wall was completed, it looked very different from the barrier of grey stone that now seems to blend into the muted colours of the uplands. Hadrian's Wall was originally white. Once the stonework was completed, masons covered it with a lime-based plaster thickened with hemp. That made an astonishing, highly visible belly-hollowing statement in the landscape. The gleaming ramparts could be seen for many miles as they snaked along the ridges and sills and across the river valleys. Rome had divided the holy island of Britain and the barbarians were to be left in no doubt about the magnitude of the power of the Empire.

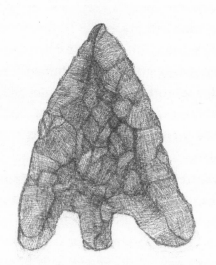

March

1 March

This is not only traditionally the first day of spring, it also feels like it. Cold and damp and dripping, Maidie and I walk out into another morning of mist. Pockets lie in hollows like patches of grey snow and gossamer scarves of mist whisper across the undulating fields north of Brownmoor. I noticed that the sucker shoots around the sycamore stumps in the Top Wood were budding and the dampness will persuade them to swell. With fewer deer to browse them, the suckers might grow tall enough this summer to be safe and become coppices.

3 March

After only a few hours' fitful sleep and my head thick with a cold, I felt full of energy. It seemed like a good day to begin work in earnest on the year's gardening. In the old conservatory converted into a greenhouse, I planted the seeds of two varieties of tomatoes in propagation trays and gave them a good soak. Then I cut down all of last year's tomato plants and threw them on the muck heap.

At the garden centre I bought four big bags of tomato compost. Once the propagated plants are potted on, I will move them to the bags and train their stems up cords attached to the roof beam of the conservatory. That at least worked well over the winter. For my outdoor raised beds, I bought Arran Pilot first earlies for one and Charlotte second earlies for the other. I grew spuds in both beds last year, so I should probably begin to rotate with other crops, like parsnips or carrots, but I have heaved in so much new muck and compost, and will add the tilthed molehill soil, that I think all will be well.

I set out the seed potatoes in trays in the conservatory so that they chit, producing tuber shoots of at least an inch before I can plant them and use my dad's old dibble. I found it in the Wood Barn at the bottom of a basket that had not been looked at for many years. His old wooden-hafted hammer was there, the tool he used to call 'the Persuader'. I also have packets of carrot seeds, some lettuce, peppers and courgettes to propagate, but I will need to buy some more trays.

4 March

Last night thick flurries of snow blew in off the hills on a snell west wind and blanketed the land under a waxing moon. Split logs from the dwindling woodpile made the fire spark and crackle as draughts found their way through the crannies of the building and warm rugs were pulled tighter around the shoulders of those who stared into the yellow flames.

Rumours had been repeated all winter, each titbit of news refreshed by travellers on the great road that led from the White Wall and threaded through the Cheviot Hills. Last summer the kindreds of the north had raided deep into Britannia, killing, raping and burning what they could not carry off in their skin boats. Called Picti, these painted warriors ignored the ramparts and the garrison of the White Wall and sailed around it, their light and fast seagoing curraghs easily outrunning the galleys of the British Fleet or sailing in shallows where Roman captains would have run aground. Like summer wasps, they flew from their nests in the north and stung the southern villas and towns again and again. Merchants from the villages outside the gates of the wall forts spoke of another invasion. Stores were being stockpiled and the southern legions would march north in the spring.

As the snow fell and the Selgovan families huddled around their fires, sleeping fitfully, sometimes shivering awake as the wind whistled down the Ettrick Valley, an air of uncertainty swirled around the thatch of their roundhouses. For many generations, the fort at Oakwood had slowly decayed. The timber towers had been fired as soon as the legions marched south in 105, and all of the usable beams of the barracks' blocks, the granaries and the principia building had been long robbed out by native scavengers. The rains and snows of a hundred winters had tumbled the turf walls until they were little more than mounds behind shallow ditches. But would the soldiers they knew as *Y Rhufeiniwr*, the Romans, march back up the valley and rebuild?

With the death of Hadrian in 138, the native kings had heard that Antoninus Pius had succeeded to the imperial purple. Anxious for military success to bolster and legitimise his accession, '[he] defeated the Britons through the actions of the governor [of the province of Britannia], Lollius Urbicus, and, driving off the barbarians, built another wall of turf'. The great depot at Trimontium was rebuilt and military traffic moved constantly up and down the country from the new wall to army command north at York. With the construction of the Antonine Wall between the Firths of Clyde and Forth, the people of our little valley found themselves inside the Roman Empire.

At first light, Maidie and I went out to find that last night's heavy snowfall had mostly melted, although behind Oakwood Fort the hills were still white.

5 March

Rain streaked with sleet slanted across the valley, driven by a fresh west wind, as I shivered in the half-light of early morning, out with the dogs to let them pee. None of them took long. After some unnaturally balmy days in late February, winter had swept back. Fortified by breakfast, Maidie and I walked out into the sleet, well waterproofed, the little dog wearing a snug coat. It struck me that we only went out in this filthy weather because we knew we could dry off when we returned. A dog walk was desirable, but scarcely essential. Until the very recent past, before inexpensive and effective waterproofs and central heating were available, no one with any sense would have gone out on such a morning.

All that the people of the roundhouses had were animal pelts worn skin-side out and slathered with fat or resin to keep out the worst of the rain, at least for a while. They became known as oilskins. Woollen cloaks, tunics and leggings woven from combed wool will have retained some of the natural lanolin from the sheep or goats but that was a fragile and temporary water-proofing. In a short time, they would have been soaked through, heavy with cold rain.

If it was absolutely necessary to go out in bad weather, then all that would dry clothes afterwards was the fire blazing in the central hearth of a roundhouse. And that will have taken a long time. It is very unlikely that our ancestors owned much of a wardrobe and damp clothes were almost certainly the main reason for the widespread incidence of acute and early onset arthritis.

Roundhouses could be snug enough if they had been well built. There were no windows and the main source of draughts was the door, usually placed in the east so that a morning sun might penetrate the gloom. Around the central hearth, the arrangement of space was radial, like the spokes of a wheel. The conical shape of the thatched roof had no space for a chimney. Where the ring of roof timbers met and were secured together, in the manner of a large tipi, any hole would have let in rain and snow, and so smoke from the fire had to seep out through the thatch. Especially on windless days this created an eye-watering interior, and in order that occupants could avoid the fug, seating around the hearth would have been low or non-existent. But the smoke that filled the upper part of the conical roof space had an important safety effect. As sparks flew upwards from the fire, the lack of carbon dioxide meant that they were extinguished before they could reach the thatch.

In 1159 BC, a volcano in Iceland blew itself apart and a vast tonnage of dust and ash rocketed into the atmosphere. The eruption of Hekla changed the weather radically, as the sun was screened for several summers and cultivation collapsed. Pollen samples and dendrochronology show a run of very poor growing seasons, especially in the north-west of Scotland and northern Ireland. How far south the effect of Hekla reached is uncertain, but some historians believe that there was famine and disruptive migration, as agriculture failed across a wide area, and that the effects lasted for many centuries. The impression is that dark times descended in the first millennium BC. It may be that the Romans invaded a land only recently recovered from a destructive spasm of climate change.

When we moved from the city to our farm twenty years ago, our attitudes to the weather altered. Instead of wondering whether or not we would need an umbrella or a warm coat, we now study the forecast each morning, using more than one website, and plan for the conditions to come. It can be tough to be outside in severe weather but at least we know we can get dry.

6 March

It has rained continuously for thirty hours. The burn behind the stable yard is threatening to burst its banks and I spent a difficult hour trying to clear the tangle of dead rushes that were damming it where it flows under our western boundary fence. Using a savagely bladed pruning saw, I hacked at the dieback and pulled out a great deal from the silted bed of the burn. But it made only a marginal difference. The tracks around the farm are awash and the loch in the Tile Field looks like it might be reforming. According to all the forecasts, we will have at least some rain every day for the next week. My drains around the houses and the yard are coping, but only just, and breaks in the downpour are needed to let the volume of water wash down into the valley's streams and from there into the Ettrick to the west.

I met Walter Elliot today, and to my delight he gave me three more very beautiful flints that had been picked up near the farm by the Mason brothers. Two are edges that were used as knives. Still razor-sharp, they will have been hafted in a wooden handle so that pressure could be exerted when cutting. The third flint is a work of art, an object of accidental beauty. An arrowhead, it is about three-quarters of an inch long, with an elegantly tapered, needle-like point. I pricked my palm with it and felt that the slightest pressure would have broken the skin. 'Aye, you would feel that if it hit you on the end of an arrow,' said Walter with a smile. It is such a delicately lethal object that I want to see if it can be safely set in a pendant as a gift for Lindsay. That would mean it being passed on down the generations, its story remembered and not forgotten in a box in the attic.

Waiting for the skies to clear, the makers of these flints had little option but to stay next to their warming hearths. When forced outside by calls of nature or a need for more wood, they probably wore brogues. Derived from the Gaelic word *brogan* for shoes, their modern design remembers what our ancestors' footwear looked like. On their leather uppers, brogues have a pattern of half-recessed holes tooled on them. These were once real holes because ancient shoes were not made to keep feet dry but to protect against cuts from sharp stones or thorns. The holes were cut to let out water as they squelched along on days like this.

Angus is a harbinger of spring. During the winter he works

indoors at a smoked salmon processing plant, but when the days begin to lengthen he leaves the seasonal job at the factory to become a busy jobbing gardener. He came last weekend and set about trimming our overgrown hedges. Little more than rows of hundreds of small trees, they urgently needed attention while still dormant. Yesterday's downpour drove Angus to seek his own warming hearth, but this morning I noticed that his trimming had revealed something else of a delicate beauty. Woven from shiny blackthorn twigs of a similar thickness, a perfectly round bird's nest was cradled in a hornbeam. It was as though the small tree was holding this little basket of fertility in its cupped hands.

7 March

Two swans flew low over the ponds on the Tile Field and suddenly, tilting their necks and wide wings upwards while thrusting out their feet, they splashed down on the largest patch of rainwater, making a momentary bow wave like a speedboat. The downpour had paused and all sorts of birds had begun feasting on its watery bounty. The swans were dunking below the surface of the pond for saturated grass, drowned worms and whatever else had been washed upwards. Keeping a respectful distance on the edge were a dozen or so ducks, quacking loudly in what sounded like glee. Overhead, its wingbeats languid, its bearing aristocratic, a heron flew slowly over the Tile Field, ignoring the vulgar cacophony below.

As often after rain, the morning was fresh, and away from the excitement on the ponds it was quiet, recovering. The tracks still flowed with run-off but the woods and fields were slowly drying, releasing the earthy scent of the land into the still air. On Greenhill Heights low clouds clung to the trees like smoke.

A consolation of the wet winter weather for the people of the roundhouses was that little or no military activity tramped across the landscape, certainly no campaigning. Packhorses and mules plodded, and carts trundled up and down the road from York to the Forth. This later became known as Dere Street. It supplied the twenty-six forts on the new Antonine Wall, though even that traffic was sporadic. At the earlier fort at Vindolanda, just south of Hadrian's Wall, several large caches of notes, lists and letters were found and in one of them a Roman officer wrote that he would not travel unless he had to because *viae malae sunt*, the roads are bad.

When Maidie and I reached Windy Gates, I realised that I had left my watch on the night table. It reminded me of holidays we used to enjoy at a big house in the Western Highlands. On the southern shore of Loch Sunart, Laudale was a magnificently isolated place and I got into the habit of making it even more detached from the hurly-burly by asking our friends to give me their watches when we arrived. I hid them in a drawer and, since there were no other clocks in the house, not on the mantelpiece or in the hall, no one knew what time it was for a week. All of the others came up from London and at first they found this timelessness disconcerting. No one was sure if they were hungry, tired or if there was enough time to go out for a walk. But soon a surprising pattern emerged. People began to get up not long after sunrise, ate breakfast, went out, even in the soft Highland rain, for long walks, came back for lunch when they were hungry, snoozed or read, had drinks and dinner and went to bed when it grew dark. A daily round not unfamiliar in our little valley two thousand years ago.

8 March

A morning of sun and ice dawned, its rays glittering and flashing off the ponds in the Tile Field and the pools on the Long Track. When the door flap was pushed aside, eastern light flooded the roundhouse and its people emerged to walk over the cracking crust of frosted ground to gather logs from the woodpile, answer nature's call, shiver and look up at the sky to judge the sort of day to come. A woodpecker drummed in the woods beyond the military road.

Spring would bring bounty: nests would have eggs, lambs and calves would be born, but rumours of war would float on the warming winds. After *Y Rhufeiniwr*, the Romans, had abandoned their turf rampart in the north and retreated behind the White Wall, there had been a few years of peace. Then the kindreds of the northern mountains and the fertile straths had raided down the great road, crossed the Wall, defeated the legions and killed their general. The Romans bought peace and, in exchange for much silver, they recovered their captives.

By AD 193 Septimius Severus had established himself as undisputed emperor, the first African to wear the purple, and fifteen years later he came to army command north at York to take personal control of the campaign against the Caledonians. For four years Eboracum, York, was the centre of the Roman world, as he mustered a huge expeditionary force, more than forty thousand men, the largest army ever seen in Britain until modern times. Marching six men abreast, the infantry stretched for nearly three miles along Dere Street. Behind them the baggage train was a tail of another two miles, and the imperial party and its bodyguards and standards must have added a splash of purple at some central, well-guarded point. Protecting the flanks of this prodigious force were cavalry regiments skirting the hills on either side. These patrols will certainly have reached as far as our little valley.

When the expedition left the depot at Newstead, Trimontium, a series of huge marching camps in Lauderdale mark its slow progress. It took four days to cover the thirty-eight miles to Inveresk on the Firth of Forth. When the end of the column was leaving one camp, the advance party of surveyors was approaching the next.

9 March

Last night the years rolled back and time seemed to be fixed at a single moment. Fifty-two years ago a team of callow schoolboys ran out onto a rugby pitch in front of a large crowd to play against a Welsh Schools team and for the first time since that damp March afternoon in 1966 we all met again. Old men travelled from deep in the south of England, one flew from South Africa, and others, like me, were much closer to the hotel in Melrose where supper and some surprises waited.

I am suspicious of reunions and, until last night, had never gone to any. They can be competitive and depressing, reminders of how unkind life sometimes is and how age really does wither. But this occasion was very different. Laughter rang round the room, good stories were told and my memory was jolted repeatedly when events I had completely forgotten were recounted. I played at loosehead prop, our captain was hooker and the tighthead prop I had not seen for fifty years. We were all big, strong lads with good technique and real skill, and we dominated the opposition in those far-off, black-and-white days. I have a photograph of us taken last night and I shall print it and pin it on the wall. We are still all big lads, just not the same shape.

Thinking about the reunion this morning, it seemed to me that it worked so well for a simple reason. We had all first come together at one of life's turning points. When the team to play Wales was announced, it was the first time any of us had been picked, selected, told we were good enough at something to represent our part of Britain. And because rugby is a team game, where players physically support each other and attempt to overcome the opposition, there was a remembered closeness that reached across a lifetime to bind us together again. I am certain that my views – political, social and otherwise – will not be shared by many of those who came, but none of this was discussed. It did not matter.

10 **March**

At first light, it was snowing heavily, big flakes falling gently out of a windless sky, eddying and swaying, blanketing the land. The old oak and the twisted, gnarled thorns by the side of the burn were mantled white, snow piling impossibly high, stacked along the branches, layer upon layer settling and freezing. Inside, in the circle of firelight, there was quiet, a hypnotised silence, staring at the yellow flicker as flames licked and crackled, the bark of a damp log sometimes hissing. On the flat cooking stone, set inside the circular hearth, sitting amongst embers, a pot seethed. The last of the winter store of barley was eked out with bones whose goodness had been long boiled out of them. Dried silverweed roots, dried funghi and hazelnut paste thickened the meagre mixture but it had at least the merit of being warm.

To pass the long hours when nothing could be done in the white landscape beyond the door, tales were told. Family stories, childish escapades, hunting lore, lambing, calving and the wider world were all woven in the circle of firelight.

In the year of the Great Army, when the first squadrons of mounted soldiers rode into the little valley, they would have found no one – deserted farms, empty barns and houses whose thatch had been pulled down so that they could not be burned. With all they could carry, and having hidden or buried what they could not, the families had driven their flocks and herds up to the high shielings in the western hills. Not daring to light a fire in the black-dark landscape for fear of Roman patrols finding them, they shivered through the long nights, whispering, listening for hoofbeats, the echo of shouted orders, the jingle of harness.

But at least their beasts would have survived, not taken because they were widely dispersed across the high plateaux. Armies only marched in the summer, when the grass grew and their horses, pack and traction animals could graze, and that was when the high pasture would have been eaten anyway, even if war had not burst over them. These dark and hard times would have made for vivid memories.

By midday at least six inches of snow lay on our fields and I crunched through it on my way to the Wood Barn. I had lit both of the woodburners at first light and they were consuming logs quickly. Nothing could be done outside once the horses in the outbye had been fed and, after a day of paperwork, we sat around the crackling fire, the draughty old house at last warm, insulated by the snow, with no wind whistling around its walls searching for the gaps in the window frames.

Perhaps in an unconscious effort to keep their memories alive, Lindsay and I sometimes talk about the stories our parents told us of their lives before we were born. Only we know them now, and before we die we must pass them on. These memories are the best sort of history, personal and vivid.

They all lived through momentous times: the Great Depression, the hunger marches and unemployment, the rise of fascism in Europe. Both of our fathers fought in the war in Europe, the Near East and Africa. My dad's medals are in a box near our bed and I shall not only pass these on to my children but also explain what they mean. In the Borders, my mum worked in the Hawick textile mills, turning out kit for the armed forces. Her neighbours came together in a tight-knit, supportive community and I remember stories of Uncle Bill catching rabbits with his ferrets, and another neighbour, an expert poacher, who gaffed salmon in the Teviot and Tweed. They kept Allars Crescent supplied with class one protein throughout the hungry years of war. Sharing, mending, making do – and at the same time coping with long spells of ignorance of what was happening to their men thousands of miles away, where battles raged through ravaged cities and countryside, and many died.

For Lindsay's mother, the war came directly to her. Working in the War Office, she experienced the terror of the London Blitz, fire-watching on the roofs of high buildings in the burning city. Not knowing if she would survive the nightly bombing raids, working long hours, she lived a provisional, exhilarating life, making fast friendships that endured until she died. She formed a friendship with a Belgian airman and wrote to him with encouragement and support as he flew sorties in the summer skies of the Battle of Britain. He was later killed in action. Helen was never so animated as when she talked about those years in the eye of the storm of war. They all lived high-definition lives that might have been cut short in a moment.

Beyond the circle of firelight, the ring of memory, the snow was still falling. When I took the dogs out before we went to bed, a pale moon lit the white land and all was silence.

11 March

A reluctant day in the city. Even though Edinburgh is spectacularly beautiful and much less hemmed-in than the canyons of central London, I find myself increasingly relieved to be going back to the peace of the farm at the end of the day. Cities are places of edges, sharp angles with no give or growth in them. And everybody seems constantly to be going somewhere, checking their phones, watching the time, waiting for the green man, busy, busy, busy.

At Waverley station some of the shops on the concourse were advertising Easter, apparently now a festival of chocolate. The vividly coloured boxes of eggs were stacked high and baskets of small, foil-covered versions were on each counter, an easy addition to a lunchtime sandwich. Like many of the rituals surrounding Christmas, all of these gaudy displays suggest recent invention, but in fact Easter eggs are an ancient, attractive tradition. Jacob, one of the Brothers Grimm, was also a folklorist and he reckoned that eggs were associated with the pagan goddess of springtime and fertility, Eostre. Easter derives from her name. Christian communities in the Middle East adapted the tradition and at the time when Christ's death and resurrection were commemorated, eggs painted red to symbolise God's blood were exchanged.

Wondering what the date of Easter was, I remembered that of course it was a famously movable feast, unlike Christmas. The Book of Common Prayer contains the formula. Easter Day is the first Sunday after the first full moon on or after 21 March. It is a very old-fashioned way of reckoning time, and that sentence probably needs to be read twice. Even more complication sets in when the formula has to be adapted to deal with clashes. If the first full moon following 21 March is on a Sunday, then Easter Day falls on the Sunday after that. This was agreed to avoid clashes with the traditional date of the Jewish Passover. In most Mediterranean languages 'Easter' is a derivation of the Hebrew word *Pesach* for Passover; *Pacques* in French, *Pascua* in Spanish, and even *Pasg* in Welsh and *Caisg* in Scots Gaelic.

The whole of the rest of the Christian calendar flows from the date of Easter and that ancient way of reckoning the land-marks of the year is very attractive. In the Borders, it is still used to work out the dates of the most important secular festivals, the annual common ridings. Selkirk's riding of the marches of the common land falls on the first Friday after the second Monday in June. Another sentence that needs to be read twice. It means that if this year the common riding falls on 14 June, then the dates of all the other towns' festivals will be reckoned in the same way.

The snow is disappearing fast, melting in shiny rivulets running down every incline. What my grannie would have called a glushie day.

12 March

A welcome day at home. Perhaps I am becoming more reclusive, or at least antisocial, but I do enjoy the rhythms of life on our farm and seem to waste less time, principally through avoiding travelling. I was able to walk off some arthritic soreness with Maidie this morning and reflected that, unlike my ancestors, it was not cold and dampness that stiffened my bones but age. The pain can sometimes be persistent and it does bore a hole in your head. I will have to start experimenting with therapies, leaving painkillers as a last resort.

As we wandered down the Long Track on a dreich, grey morning after heavy overnight rain, I noticed scores of worms on the hard surface. With their burrows filled with water, they had been forced to come up to the surface to breath. Exposed on the track, I suspected they would soon be food for the hungry crows.

13 March

Fierce winds raged around the house, roaring and buffeting, rattling the slates. *Fiadhaich* is a word the Gaels use, and it means savage, like a wild beast. The Hebridean crofters built their black-houses with massive stones and no windows so that they could keep out the worst of the savage winds that whipped off the Atlantic, and their thatched roofs were held in place with *simmens*, lattice works of heather ropes weighed down with big stones. These are ancient and were probably used to keep the roofs on roundhouses. Severe storms must have wrought great damage and literally blown houses down. Only in the modern era have we been able to build largely reliable shelter. In the long past, the threat of winds like last night's was very real.

16 March

More late winter snow fell overnight and long into the morning. Much of it melted on the warming spring ground and I suspect that by the end of the day it will have gone, leaving the fields quagmire deep and the tracks running like torrents. As the land blanketed white, I watered my tomatoes in the conservatory, a surprising contrast. The Italians call them *pomodori*, golden apples, not after the sun but because the early varieties brought from South America were yellow. I planted three different varieties. San Marzano has some yellow colouring, Ildi is a cherry tomato that grows in a bush shape and Red Pear is the misleading name of a plum tomato. They are sprouting nicely and I hope they do better than last year's.

When I plodded down the Long Track with Maidie, three tall clumps of marsh grass in the Tile Field began to run towards the fence. The camouflage of the little roe deer, especially in the grey light, is almost perfect.

17 March

Apart from where it had drifted a little along the fence lines, yesterday's snow has disappeared and a brilliant morning sun lit the dripping, sodden land. It was time for Maidie and I to climb. Dominating our farm is the hill of the Deer Park. Fifty-six acres of rough grazing pock-marked with three small quarries where the whinstone that made the farmhouse was hacked out, it is a magical, spectacular place. Stories are buried everywhere. From the highest point, the vistas to the west are ten miles long, reaching up the Ettrick and Yarrow Valleys, and to the east, the site of the Roman signal station on Eildon Hill North can be made out.

In the twelfth century, the park was created by royal deerherds who served the macMalcolm kings. David I had a castle built at Selkirk and the mounds of its motte-and-bailey lie just outside the north-west corner of the park. Raised at the court of Henry I of England, David was a progressive, modernising monarch, a break with the Celtic past, and when he succeeded to the throne of Scotland in 1124 he brought new ideas and new people to the north. Like the English kings and their nobility, David was addicted to hunting and he appropriated the lands of the Ettrick Forest as a royal game reserve. It was vast, extending westwards from his new castle for hundreds of square miles. In the Middle Ages, a forest was not so much a large wood as an area of wild, uncultivated land where game could breed and run free.

When Maidie and I climb to the summit of the Deer Park, we can see long tracts of the old hunting forest, but today we stood at its base, the place from where it was managed. Around the bounds of the park are long runs of the original ditching, with a bank made from the piled upcast on the far side. Into the bank were rammed *pali*, the Latin word for stakes, and these could be six or seven feet high, with the ditch in front making them even higher. The word *palus* morphed into *pale* in English, for a demarcated enclosure, and is the origin of the unpleasant phrase 'beyond the pale', for those on the outside.

Frightened deer can jump very high and the fence was built to keep in pregnant does through the winter and keep out predators such as wolf packs. The deerherds fed the does forage, so that they could survive to safely drop their fawns in the spring. Then they were all released back into the forest to roam free and be hunted by human predators and their hounds. It was an early example of gamekeeping, and on a grand scale.

Where it runs uphill from the East Meadow, the ditch in front of the park pale is cut through rock, a great labour that struck me as unnecessary. Within ten yards, the topsoil is much deeper and our recently planted trees are thriving in it, sending down long taproots. Why did the ditch have to run on exactly that line when the king's authority could have ensured it was placed almost anywhere? I wondered if the labourers who hacked through the very hard limestone had been marking a much older boundary, the margins of an earlier lordship of some sort. It was a question that would be answered by a remarkable discovery.

By the fifth century, the Roman Empire in the west had faded and, after repeated barbarian onslaughts, it finally disintegrated. The garrisons of the White Wall were long gone, even though much of it remained. The long middle sections run through bleak moorland and the cut blocks worked by the Roman army's masons were less easily accessible to stone robbers. While its white plaster had crumbled after the snows and rains of hundreds of winters, the fabric of the wall endured, a monument to faded power. Into the vacuum came many people, their movements, expeditions and settlement rarely recorded. In the centuries after Rome, the Angles, Saxons and other Germanic peoples sailed the North Sea and created kingdoms in the south and east of Britain. From the west came the Irish warbands. In Argyll, they carved out the kingdom of Dalriada, brought the Gaelic language, contended for, and in the ninth century won, the kingship of Alba, a large part of what is now Scotland.

Language and cultural shifts are processes, not events, and therefore very difficult to date and trace. Place names are suggestive and sometimes the incursions of raiders were noted by monastic chroniclers, principally Bede of Jarrow. In 603, a Gaelic-speaking king of Dalriada, Aedan macGabrain, was defeated at Addinston in upper Lauderdale, about twenty miles north-west of our farm. On that bloody day in the eastern hills, the victor was Aethelfrith, king of the Angles, and in a series of lightning strikes in little more than a decade he and his sons extended their reach over the Tweed Valley and beyond. But such victories are rarely decisive or complete and I wondered if Anglian control extended as far as our little valley. One blustery morning, walking in the Deer Park, I discovered that it did not.

Sheepwalks shelve around the flanks of the park. Narrow little paths trodden and indented by the small hooves of generations of ewes, their lambs and tups, they can be difficult to walk but dogs like them since ancient scents seem to survive. On a sheep-walk below me – for some reason there are often several at different levels going in exactly the same direction, making the hillside look like crimped hair – I noticed a large stone about the size of a square football, so to speak. It had been worked and had a flat top. The underside had been smoothed over a long exposure to the elements, but it was irregular and it looked to my inexpert eye that it had been broken off, perhaps having formed the top of a pillar of some kind. But more striking were a series of markings on either side of one of the straight edges. In a moment, I realised what I had found. It was part of an inscription, not in Latin, but in the tree language.

Several years before, I had become interested in Ogham. This was a rune-like language that originated in Ireland and spread to the west of Britain in the centuries after the fall of Rome. Using the long edge of a piece of stone (trees were almost certainly used much more widely, but of course they have not survived) to represent the trunk of a tree, chisel-cuts were made in straight lines on either side or sometimes through it. Known as *fleascan* or twigs, these were the letters of the Ogham alphabet. For example, A is a straight line through the trunk, R is four diagonal lines through it, and H is a straight line but only marked to the left. Turning over the squared-off stone in my hands, I could make out two letters, an A and an H.

Complete Ogham stones usually carry personal names and were probably boundary markers. My stone had been broken and I set off down the steep slope, crossing the sheepwalks, to look at the drystane dyke at the bottom. Perhaps the rest of it had been picked up by dykers and used as handy, squared-off stones to make a flat bed. But even after several careful searches over a week, I could find nothing.

The top of the Ogham I had was nevertheless eloquent. Reasoning that it had fallen from the summit plateau of the Deer Park, I looked for where it might have been set up, and one of the most likely and visible sites was near the run of rock-cut ditching that had formed part of the park pale raised in the twelfth century. If I was right, then I had found solid evidence for something I had long intuited, that the story of this farm is indeed a palimpsest, history piled on top of itself, layered like sediment.

The chance nature of my find was at the vanishingly small end of any scale of likelihood. To find the stone was one thing, but to recognise its inscription was quite another. I contacted one of the few other people in Britain who can read Ogham and agreed to bring what I had found to Glasgow University. When I pulled the stone out of the plastic supermarket bag that had sat in the luggage rack of the train, the eminent historian's eyes widened. She stood up, walked around her desk, looked closely at the stone, looked at me and took the silk scarf from her neck and for some reason wrapped it around this little piece of history.

No Ogham had been found as far east as our Deer Park and its presence suggests a patchwork of settlement in the time known as the Dark Ages. Despite Aethelfrith's victory at Addinston, Gaelic-speaking chieftains did penetrate the upper Tweed Valley and they appear to have created a boundary of some kind between them and their eastern neighbours, most likely Angles. One of them had a name that began with A and H. As a finder of so-called treasure trove, I was sent a cheque for £75 by someone called The Queen's Remembrancer and my square football is now in the Royal Museum of Scotland. Perhaps one day Maidie and I will find the rest of it.

18 March

A gang of honking geese are making themselves heard amongst the ponds in the Tile Field. Marching this way and that on the sodden grass, chests puffed out, necks craned, they seem like loud, waistcoated boozers with too much beer in them at a rugby match. Even the gulls are staying out of their way. More sun warmed the morning but springs still bubble up everywhere. Finding their way through a stony, subterranean network, pulled on by gravity and pressure, these unceasing trickles wash the silt down the Bottom Track in such quantity that I have had to clear the drain gratings twice in the last week. In the intervals between the honking chorus in the Tile Field, I heard the sweet, fluting song of a bird perched somewhere in the trackside birches. I could not see it. How I wish I could recognise these calls.

19 March

'Same old, same old' was the reply when I asked an old friend how life was. The sense of repetition saddened me, but perhaps I could have said something similar. Each morning I follow the exact same routine with the dogs, feeding them, taking them out for a pee and a poop, taking Maidie up the track. Yet I don't feel that sense of repetition. Maybe it is because I am outside every morning, noticing the weather because I have to, listening to the birds, watching the seasons change, even if the processes are slow. Change is constant.

This morning felt spring-like, patches and strips of new grass were flushing in the fields, and when our neighbour had finished delivering hay to the Old Boys in the East Meadow he stopped the tractor to exchange news. The lambing has started and the cows that have not yet calved are roaring each morning to get out of the byre and into the fresh, warming air.

21 March

Faint on the west wind blowing over the hills from Whithorn, God was singing. For those who lifted their heads to the heavens and could hear him, those men and women believing in Christ, Ninian's priests walked through the wild land to show them the shining path to glory. So that God could see them as their mortal bodies were laid in the cold ground in the sure and certain hope of resurrection, stones were carved and set up. Carantius, son of Cupitanus, Coninia, and the princes Nudus and Dumnogenus were servants of the Lord and their people commemorated them in the sacred language of the church.

At Luguvalium, modern Carlisle, a Christian church flourished in the Roman city and endured beyond the end of the province of Britannia in 410. Ninian was almost certainly a member of that community when he was sent to Whithorn in Galloway to build a church 'after the Roman manner' – that is, out of stone and not wood. Perhaps because of its colour, or its sanctity, it was called Candida Casa, the Shining White House. From there, missionaries walked over the watershed hills to the east and faint traces of their passing have been found.

Inscribed stones mark the steady progress of the word of God, the Latin of Holy Mother Church. Near where the Ralton Burn flows into the Liddel Water, about a mile north of Newcastleton, a drystane dyke was undercut by flooding and its stones toppled into the water. One was very large, nearly six feet long, and those who came to rebuild the dyke noticed writing carved on its surface. It read *Hic Iacet Caranti Filii Cupitani*, Here Lies Carantius, Son of Cupitanus, and archaeologists dated the stone to the fifth or sixth century. The *hic iacet* formula indicates a Christian burial.

In 1890 a slab of hard whinstone was pulled out of a small cairn in the beautiful Manor Valley near Peebles. On it were carved two words and a small cross. It was the headstone of Coninia and next to it was a cross dedicated to St Gorgian. He was an obscure Syrian martyr who died around 350 and the medieval parish church of Manor was later dedicated to him. These are gossamer traces of a forgotten sanctity, a memory of a saint whose name and fame have long faded into the darkness of the past.

Most striking was an ancient memorial unearthed in the Yarrow Valley, only seven miles west of our farm. In 1803 a ploughman dug up a large flat stone with a Latin inscription on one side. It was taken to Bowhill House, the home of the Duke of Buccleuch, and examined by him, Walter Scott, John Leyden and Mungo Park, three of the most famous Borderers who ever lived. The inscription is very mysterious, for it speaks not only of a belief in Christ but also of a vanished society and its leaders:

> THIS IS AN EVERLASTING MEMORIAL.
> IN THIS PLACE LIE THE MOST FAMOUS PRINCES,
> NUDUS AND DUMNOGENUS.
> IN THIS TOMB LIE THE TWO SONS OF LIBERALIS.

This royal sepulchre is fascinating. It remembers a dynasty who ruled in the centuries after the legions marched away, and who gave themselves Latin names and who worshipped God. Who were these men? Liberalis was probably a king, and his name might translate as the Generous One, possibly a cognomen, certainly a signifier of a necessary virtue of post-Roman British royalty. The scatter of bardic sources in Old Welsh, the language spoken by these three men and their people, sing of kings who gave gold and horses to ensure the loyalty of their warbands. Here is a fragment composed by the great bard Taliesin, about the king of Rheged, a realm that may have encompassed both shores of the Solway Firth in the sixth and seventh centuries:

> URIEN OF ECHWYD, MOST LIBERAL OF
> CHRISTIAN MEN,
> MUCH DO YOU GIVE TO MEN IN THIS WORLD,
> AS YOU GATHER, SO YOU DISPENSE.

Like the king commemorated on the Yarrow Stone, Urien too was *liberalis*, generous. The native name of one of the famous princes may have been Nudd, but the other, Dumnogenus, is mysterious. It may mean something like 'born in the world', but that literal translation does not take us any further.

Archaeologists have dated the Yarrow Stone to around 500 and such scant sources as exist talk of a kingdom that may have lain between Rheged in the west and Calchfynydd in the east, in the Tweed Valley. It was known as Goddeu, the Trees or the Forest, and this Old Welsh name may be remembered in Cadzow (which used to be pronounced Cadyow, the substitution of the z being a later mistake), a village on the upper reaches of the River Clyde. Goddeu is a Janus name, looking backwards as well as forwards. The hill country of the Upper Clyde, Tweed and Annan was once the kingdom of the Selgovae, and it later became the vast hunting reserve of the Ettrick Forest. It sounds a surprising echo on winter Saturday afternoons. When I occasionally go to watch Selkirk play rugby, some supporters shout an ancient encouragement from the terraces: 'Hawway the Lads o' the Forest'.

When the Yarrow Stone was returned to the place where it was unearthed, around it were found the remains of an early sixth-century cemetery, probably Christian because of the east/ west orientation of the graves, and in the Roman habit it had been dug next to an ancient track. The stone stands near the centre of a wide panorama, on a plateau surrounded by sheltering hills of three sides. It feels like a good setting for the tomb of the princes of a forgotten kingdom. But all that is certain about Nudus and Dumnogenus was that God was in their hearts.

22 March

The yellow is on the gorse at last, luminous even on a grey, blustery day. Like the daffodils waving in the wind, the bushes on the flanks of the Deer Park wear the rich yellow colours of fertility, rich and warming. Walking down the track into the park with Maidie, I noticed what looked like some much decayed stonework tangled in the roots of a sitka spruce in the wood next to the Haining Loch. Summer leaves must have hidden it on the countless times I had passed it.

Once I had negotiated the barbed-wire fence to get closer, I could see that it was indeed stonework and that it had been skilfully done, even though much of it had been pushed around by the roots of the tree. There were three or four clear courses of what felt like sandstone and each block was of a uniform size, a little larger than a modern brick. All of the outward faces had been chiselled smooth. It seemed that the bank had once been much higher and I surmised that it had formed part of the western run of the park pale. Nearby is a place where the Long Track curves away around a hillside and old estate maps mark it as Windy Gates.

I wondered if I had come across a deer leap. In the early winter, deerherds took their dogs into the Ettrick Forest to drive pregnant does towards the haven of the Deer Park. Wide entrances were made in the pale and the does were herded towards them, the hounds funnelling them tighter and tighter. From the forest side, gaps in the fence looked like a means of escape, but in fact they were deliberately deceptive. Like a very high ha-ha wall, there was a steep drop on the far side of these wide gates, too steep for the does to jump back up. The wall needed to be sheer, smooth and stable, and I wondered if the well-built stone revetting I had found was the foot of a deer leap preserved in the shadows of the wood by the loch.

23 March

A very shaggy, winter-coated Maidie went to the dog groomer in Selkirk and she came back shorn and shelpit, a Scots word for skinny. Even though we had sun this morning, I put the wee coat on her as she sniffed after bunnies.

24 March

Lately I have found my professional life frustrating, difficult, and I thought I might drive over to the Holy Island of Lindisfarne for a break, a few hours to myself. I love that place, and I decided some time ago that if I felt I needed it I should seek its peace.

On a bright, blustery Sunday morning, I found myself summoned by bells as soon as I arrived. St Mary's Church is ancient, parts of it Anglian, and it stands next to the ruins of a later Benedictine priory. A warm and welcoming church, with the sun streaming through the stained glass casting a kaleidoscope of colour, its congregation of thirty-five or so, most of them probably islanders, had gathered and the organ was playing. A woman smiled at me, came across to my pew and asked if I was Graham, a vicar from a church on the Wirral who had come on a retreat. It made me smile. I have been seen as and called many things but never before a vicar. As the service began (with the vicar of St Mary's asking if anyone needed a gluten-free communion wafer), I left to go and walk around the island.

At the farthest north-east corner stands the day-mark at Emmanuel Head, a white obelisk raised to discourage shipping from coming too close to the offshore reefs. As usual, the crowds of visitors had stayed in the village or gone to visit the fairytale castle, and at last I had some solitude, but no real peace. The island is tidal and the causeway to the mainland would become impassable at 2.45 p.m. That kept me checking my watch, forcing me to follow a timetable.

But I did go to the places I liked and I had time to fulfil one important vow. In the cemetery around St Mary's I had come across a headstone with a heartbreaking inscription that had brought tears when I saw it last year. Hugo, a baby of only fourteen months, had died and I wept for him and for the hurt suffered by his parents. In 2015, Hannah, our first granddaughter, died and I had some sense of the weight of their sadness. In St Mary's I lit candles for the wee ones, and another for Grace, my sparkling granddaughter, Hannah's little sister.

Before I had to catch the tide, I climbed up to the Heugh, a narrow rock that rises to the south of the priory ruins and looks out to sea. I was alone there with my thoughts for a while. It was a good moment.

26 March

Three oystercatchers were wheeling and wheeping in the grass park to the west of the Long Track. Usually coastal birds, we sometimes see them in spring when they come inland, poking and strutting across the damp, dewy grass. Their flight looks more like display as they rise up, banking, tilting their wing-spreads, showing their white undersides and wingbars. Their piping wheep-wheep call is one of the sounds of spring.

Around the farmhouse flew more harbingers of the warm seasons to come. The little wagtails have arrived, bobbing around the outbuildings, their tails wagging up and down, their beaks full of horse hair or small twigs, quickly taking off in their undulating flight that seems somehow to be jumping fences.

27 March

Looking after my granddaughter for a short time today was easy. Probably because she had a cold, Grace slept deeply in the middle of her parents' bed. On her back, with her head turned to one side, she looked perfect, angelic, completely at peace.

28 March

It was a still, black, moonless night and the stars were hidden behind dense cloud. Suddenly the silence was rent by frantic thrashing in the pinewoods north of the farmhouse, by unseen movement, by something big, angry and uncaring of any noise it made. I stopped dead at the foot of the Bottom Track, peering into the blackness, my little dog beside me, frozen, not barking, her ears pricked and tail up. At any moment I expected something, no idea what, to explode from the trees. The most dangerous predator to stalk the countryside is man.

The thrashing paused. And then moments later I saw a shape race across the track and behind the trees that fringe it. Shapes in the night can be made into anything and I had no idea what I had just seen, except that it moved too quickly for a human being. All I knew for certain was that it scared me.

Dylan Thomas wrote of nights such as these as bible-black, but I think they are pagan-black. Before the twentieth century brought electricity to even the remotest country districts – as well as lighting streets, cars and buildings – dark nights were completely and unrelievedly dark. Only oil lamps could be taken outside and in any sort of wind they could be quickly snuffed out. When the world was black-dark, few ventured over the door unless they had to. The world of the night was dangerous, for who knew what lurked in the woods, or rose up out of the mosses and sikes to grip an ankle and pull down the unwary? What dancing lights led the lost to flounder in the sucking morass and draw them down, spluttering and choking, into the cloying deeps of the world?

Writing this in the bright morning after, these fears seem like melodrama, but last night, for more than a moment, I confess I was badly frightened, and I still have no idea what I heard and may have seen.

29 March

Having propagated rows of seedlings in trays, I potted on the most vigorous shoots into small fibre containers, ones that will disintegrate when I plant them in Gro-Bags in the conservatory. I planted more seeds than I needed and it always vexes me to throw away the little seedlings I don't want. It is not only the waste; the extinguishing of green, delicate, spring-like life does upset me. I don't like killing any living thing, but it is ridiculous to be so absurdly sentimental about plants. And yet I can't help it. Perhaps it is because I grew them from seed.

30 March

The beginnings of blossom are peeping through the hedges and woods. No scents yet but they will come, and a first fresh flush of leaves is unfurling on the geans and thorns.

31 March

This is both Mothers' Day and the first day of British Summer Time, what feels like a fitting conjunction. But even though the clocks moved forward and I had an hour less sleep, it does not matter. Clocks are irrelevant here. What matters is that I saw a beautiful, metallic pink dawn followed by a brilliant morning sun whose warming fingers reached across the land, defining its ridges and swales. Looking over to the Deer Park, I could see the line of the ditch of the park pale, even where it had been partly obliterated by a small quarry. Maidie and I walked across the field behind the wreckage of the Top Wood, dodging fresh cowpats as though they were landmines, and we stood on the highest point of the ridge to look to the west over our little valley and to the Ettrick Hills beyond. Creation looked well this morning.

April

1 April

Huntigowk! Huntigowk! Meaning 'hunt the cuckoo', it was what we used shout in our street at someone who had been caught out on April Fools' Day. It is a tradition that seems, sadly, to have waned. Perhaps there will be a Loch Ness monster story in the papers.

On another crystal morning, cold and very brilliantly sunny, it felt instead like Groundhog Day, an American import but an original one. Yesterday I saw a solitary heron fly high over the Long Track and the Tile Field before disappearing into the distance over Brownmoor. This morning at exactly the same time a heron flew in exactly the same direction. Let's see what happens tomorrow.

Out before dawn to pee the dogs, I looked out to the west. Moving slowly across its flank, a flock of sheep climbed Howden Hill to catch the first rays of the sun. I don't know why but it was one of the most beautiful sights I have witnessed so far this year.

2 April

Yesterday saw another of the year's turning moments as the first of the spring lambs came out to grass from the warmth of the heat lamps and the straw in the shed where they were born. They were in the park to the east of the Long Track. Four ewes herded six lambs, one of them jet black, to the fence line below the Young Wood. Not only was there shelter from a stiffening breeze, the new grass had been flushing near the warmth of the trees. It was a sight to lift a winter heart.

3 April

A chill wind was blowing out of the north this morning and the ewes and lambs were once again hugging the fence line by the Young Wood. It runs east to west and its overhanging boughs sheltered the wee ones. I could not see any lying in the sodden grass and so their mothers' milk must be sustaining them through the bad weather. There was snow on the ridge where the Thief Road runs and I remembered that the hill lambs are always later.

4 April

Two bewildered worms and a snail slithered slowly across the snow on the Long Track. It had been falling since the early hours, big flakes floating down and covering the land quickly. In the grass park beyond the track the ewes and lambs would shiver in this late winter flourish and have to scuff the snow aside to find grass, but it is better than wind-driven rain. Equally important for us was the question of being able to drive up the track.

5 April

A bright day and buds are beginning to open.

7 April

For the long, sandy coasts of South Uist, winter flood tides have an immense length of fetch, their waves made by winds off the coast of Canada and driven three thousand miles across the vastness of the mighty Atlantic to spend themselves on the beaches and cliffs of the Hebrides. These titans rear up and crash with tremendous elemental power, but instead of destruction they bring life to Uist. Ground by the unceasing churn of the sea, shells become powdered into calcium-rich sand and when the winds carry it inland, it neutralises the acidity of the peat and makes the machair, a fertile grassland littoral of great beauty.

Warmed by the waters of the Gulf Stream, the machair begins to flush in April and, long dormant, many hundreds of thousands of wildflowers push through the rich soil and begin to bud. By the beginning of May, the colours are dazzling: waving in the offshore breeze are sea-pinks, purple harebells and bluebells, yellow trefoils, eyebright, gentian and the intense scent of wild thyme. Ground-nesting birds like the corncrake, chough and corn bunting call and sing, feeding on the insects while yellow bumble-bees drink the nectar of the machair's flowers.

Three thousand years ago, our ancestors grazed their little black cattle by the ocean and built houses on the margins of the grassland. At Cladh Hallan, archaeologists uncovered ancient ruins and made an amazing find under the floor. The remains of two mummified bodies emerged from the soil. At about the same time as Tutankhamun was placed in his golden sarcophagus, this Hebridean community understood that the dense, anaerobic peat would preserve corpses and, immediately after death, they took two important people and placed them deep in a bog for between six and eighteen months. On exhumation, it was found that all the processes of decay had been pre-empted. The bodies, a man and a woman, were then mummified but not buried. Perhaps they were displayed in a special building. Sometime around 1000 BC they were both buried under the floor of the house discovered at Cladh Hallan. No one can do anything but theorise about the beliefs of these people, except to note a long and very moving continuity.

Cladh Hallan is still a place of the dead. On a low mound in the machair, rows of gravestones punctuate the horizon, and this year one of them will be raised to an old friend. A native Gaelic speaker from South Uist, he died last week and tomorrow I will go to his funeral at the Catholic church in Melrose before his family take him back to home, to Cladh Hallan, where his people are waiting for him.

8 April

The air was filled with the vibrato bleating of hungry ewes. More than a hundred are now out in the grass parks, with their tiny lambs nickering behind them. From what looks to me like a golf buggy, my neighbour shakes out sacks of hard feed as the sheep gather around him. Some of the lambs look vulnerable and I saw one with a deformed foreleg. Perhaps it will straighten out as the richness of its mother's milk builds bones.

It is a grey morning. The land seems not to stir, a good day for a funeral. In my friend's memory, I shall recite a Gaelic lyric we both loved, and it will seem sad and strange to me not to be able to speak to him in the mouth-filling language of the islands.

For centuries, a dialect of Northumbrian English, what is now called Scots, has been everywhere in the Borders and Central Scotland, but it was not always so. In his magisterial *Ecclesiastical History of the English People*, Bede of Jarrow lists the five languages spoken in Britain: Latin, of course, the language of the book, the Word of God, along with English, Old Welsh, Gaelic and Pictish. Only one of these now remains, but the map is speckled with place names from all of them, including Latin. Within only a few miles of where I sit writing this is Selkirk, an Old English name that means 'the church by the hall or manor house'; Kelso is derived from Calchfynydd, Old Welsh for 'chalk hill'; Glendearg near Galashiels is from Gaelic, 'the red valley'; Eccles in Berwickshire is ultimately from the Latin word *ecclesia*, a church; and further afield, Perth is Pictish for 'copse' or 'thicket'.

The Ogham stone I found in the Deer Park was inscribed in Irish Gaelic, so there may have been an old, lost name for our farm. Language shift is impossible to trace accurately in the deep past, but I am certain that Bede, Cuthbert and the early saints lived in a multilingual society, something now sadly lost. And now, with my old friend's death, another of Scotland's old languages falls silent in the Borders.

9 April

Swinging their picks in a rhythmic arc, allowing the weight of the tine to do the work, the farmers and shepherds were digging a ditch. Once the earth and stones had been loosened, women and children shovelled it into wicker baskets and carried load after load uphill to pile it on a bank. Having tipped out the earth, they stamped it down on the upcast. On the rampart of the old fort, only a few yards uphill, warriors watched the work go on. Built many centuries before by the Old Peoples, circled by steep and deep ditches, the fort corralled ponies and was a good place to camp.

After the funeral yesterday, I parked below the Rink, a striking prehistoric hillfort close to the junction of the Tweed and the Ettrick, between Selkirk and Galashiels. Halfway up the slope below the fort is a long copse fenced off to keep out curious lambs. It is a short section of a lost frontier known as the Catrail. Probably dug sometime in the seventh century, a deep ditch and bank ran for fifty-eight miles across the valleys of the upper Tweed and Teviot before disappearing into the Cheviot Hills. It marked a cultural and, at first, a linguistic divide. To the west of the Catrail are many place names that derive from Old Welsh, and to the east, most of the farms, villages and towns are named in English.

The copse is very atmospheric. Once I had climbed the fence without mishap, feeling a little overdressed in the suit and tie worn for the funeral, it was possible to forget that the twenty-first century rumbled below on the A7 and to find myself standing at a pivotal place on the faded map of an ancient Scotland. Even though none of the stones that had tumbled into the ditch had been worked and only faint traces of this great labour remain, there is a *genius loci* in this unlikely survival of the old frontier. It is a place of spirits not yet fled.

Some of the wisps of history that float in the air around this atmospheric place are unexpected. Until its adoption by the Nazis in the 1930s transformed it into a symbol of evil, the swastika was thought to bring good luck and also be a link to the gods. For the Romans, it was a representation of the forked lightning of Jupiter's thunderbolts. Recently an iron swastika fob was found just below the line of the Catrail. A rare object, it was suspended, dangling below a part of the harness ponies wore when they pulled chariots. In the last few centuries BC, warriors of the native British kindreds were expert charioteers, able to race their ponies over uneven ground, bouncing up and down but keeping their balance and a tight hold of the long reins. In his description of the fateful battle at Mons Graupius between the Roman legions and the Pictish host in AD 83, Tacitus wrote of them moving back and forth in front of the native army. Warriors are superstitious and perhaps their swastikas were good luck charms. But what were charioteers doing at the Catrail? The chance find of this object is like a momentary, blinding flash of light in the black darkness of prehistory.

There was also a wider significance at play on either side of the Catrail. Behind the rampart lay the lands of the hillmen, shepherds and hunters, and to the east were the fertile fields of ploughmen, the valuable, productive farms and villages claimed by the Anglian invaders, the warbands of the dynasty of Aethelfrith, the warrior-king whose men had overrun the lower Tweed Valley.

The morning clouds had blown away and the sun streamed

through the bare branches of the thorn trees. The breeze had dropped and, as I walked back down the hill, I watched the lambs come to life. In the warmth, groups of them skittered unsteadily around the field, some of them jumping as though on springs. Gambolling is, I believe, the technical term. In her excellent eulogy, my friend's daughter had said how much her father had loved the springtime and that he could have lived life on a continuous loop from late March to the end of June. Perhaps the swallows he loved so much will come soon.

10 April

Swaddled in lanolin, the lambs lay on the frosted grass, waiting for the sun. After temperatures had dropped to well below zero, the morning fields were white. A stiffening breeze riffled the wind-worn willows by the Common Burn and the sun's rise seemed slow and sluggish, hidden by gauzy scarves of cloud on the horizon.

When at last Maidie and I walked down the Long Track, she kept stopping, planting her forepaws on the stones. After a few tugs and words of encouragement, I realised that she was apprehensive about passing a small group of ewes gathered by the fence. Very un-terrier-like. Their lambs lay in the winter wrack, the longer stalks of weeds and old grass on the edge of the field, and these mothers were not for moving. A dominant ewe stamped her forelegs on the frozen grass.

By the time sunshine had flooded across the landscape, it was as though a series of triggers had been pulled. Along the fence line of the Young Wood a flock of thirty or forty lambs began to race together, back and forth, while their mothers waited patiently for my neighbour to arrive with breakfast, bags of hard feed to supplement the sparse grass. The breeze began to drop and the land to settle to the rhythm of another spring day. This time in the year sees progress, sometimes halting, as the animals and their young thrive, the leaves unfurl, the birds gather nesting twigs, moss and horse hair, and the warmth returns.

12 April

For more than a thousand autumns, my ancestors barely rustled the leaves of the woods or left little more than gossamer marks on the land. The Romans were builders and road makers, the legionaries and auxiliaries experts in their various trades, their prefects and officers having all the skills of civil engineers. No series of concerted or organised efforts at building anything that endured followed their departure from Oakwood sometime after AD 100 until the medieval period. The park pale and the motte-and-bailey castle in the corner of the Deer Park were substantial projects but probably not dug until the later twelfth century.

This morning's hard frost and low, dawning sun showed up the characteristic pattern of runrig ditching across the south-western flank of the Deer Park. Between the lines of drainage ditches on what is a steep and difficult gradient, the upcast was piled to make long cultivation strips. On that sunny slope, barley was probably grown, the altitude too high for wheat. But the runrig is impossible to date. It might even be a remnant of what was known as Napoleonic ploughing, an attempt to bring into cultivation as much land as possible in the early years of the nineteenth century, when imperial France dominated Europe and isolated Britain.

13 April

Chance is indeed a very fine thing, but combined with coincidence it can sometimes seem like providence. The day after complaining about the darkness of the past, a bright light suddenly shone on the story of the Long Track that made it even more of a highway through history. Last night, I had an email from a metal detectorist who had been sweeping the northern section of the track. After rounding the little hill that my children used to call Huppanova (because the ground rose up and down over the other side), it runs due north past a ruined dovecot before swinging east towards the Georgian mansion house known as the Haining. It was the centre of the estate of which our farm used to form a part. So much earth was shifted in building and landscaping that it hid the original route of the track. In the early twelfth century when the motte-and-bailey castle in the corner of the Deer Park first comes on record, it led from its west gate. Two centuries later, the Long Track rang with the clangour of war as soldiers tramped, carts creaked and the heavy horses known as destriers snorted. Armies were marching north and the skies over the Borders darkened.

In his fascinating email, the metal detectorist listed the many medieval coins he had found and added that he had turned up part of a small, silver crucifix, perhaps worn by a woman. 'Hen's teeth rare' is how he expressed his delight at discovering a German coin minted in the name of the Holy Roman Emperor Otto IV. His troubled reign ended in 1215 after a bitter civil war fought all over Western Europe in which English kings were heavily involved. Richard I had made Otto Earl of York in 1190 because he was his brother-in-law, and later King John had sent a vast subsidy of 6,000 marks in the form of about one million silver pennies. Despite the English cash, Otto's cause failed.

Some of the other medieval coins were minted later and date from a period of intense activity at the castle and on the Long Track. After the death of Alexander III in 1286 and the ensuing succession crisis, Edward I of England used the opportunity to establish his overlordship in Scotland. There followed more than a century of intermittent warfare. Between 1301 and 1304, Selkirk Castle was rebuilt by the English and a stone keep raised on its mound. Carts trundled along the Long Track carrying stone (perhaps some of it came from the three quarries in the Deer Park), timber, lime and other materials. The Exchequer accounts recorded the immense cost of the project, as the local economy was flooded with cash. More than 329,000 silver pennies were paid to workmen and suppliers. As they made their way along the track, some of these people dropped coins that were trampled into the mud to wait seven hundred years to see the light of day once more.

What these finds mean is something striking. Unlike the runrig on the flanks of the Deer Park, the Long Track can be dated by the coins lost on its margins. Not since the flints and the spindle whorls picked up by the Mason brothers have the fields around our farm yielded such a pungent sense of their past. I walk the Long Track almost every day with Maidie, and while I know all the puddles, ruts and big stones I have little idea of the stories that lie only inches below my feet. But I feel the ghosts of the past whispering beside me and, even on sunlit mornings, their quiet tread can be heard.

A small village formed around Selkirk Castle, its earliest houses probably built along Castle Street on a ridge that runs downhill from a gate in the north-western defences. When the castle fell into disrepair and eventual disuse in the later fourteenth century, the Long Track seems to have shifted its route slightly to the north so that it arrived at the West Port, the gate to the growing village. It then began life as the main road between Selkirk and Hawick.

14 April

It is Palm Sunday. A week before Easter and the beginning of Christ's passion, it recalls his triumphal entry into Jerusalem, when crowds strewed palm fronds in front of him and his donkey. All over Christendom the faithful used to flock to their churches to pray and to celebrate. Inside the cemetery or the sacred precinct, crowds ate, drank, danced, and played music and games without inhibition, a high contrast with the aura of solemnity that surrounds churches now. In his *De Temporum Ratione* (*The Reckoning of Time*), Bede recognised that the term Easter came from Eostre, the Anglo-Saxon dawn goddess associated with the renewals of spring. Christian missionaries quickly made a link with Christ's resurrection, but the cakes and ale were probably a harking back to pagan practices.

The sun shone on this Palm Sunday morning and I took Maidie and Freydo down the Long Track. Yesterday the normally aggressive little Westie had to be dragged past the ewes and their lambs standing by the fence – outnumbered by small, white creatures that looked a bit like her, and intimidated by their mothers standing their ground, she was scared. So my wife suggested I took Freydo this morning and, after a little initial reluctance, all was well.

15 April

The archers hiding in the woods above the river watched a belly-hollowing sight. Strung out as far as the eye could see along the old Roman road on the south side of the Tweed was an enormous army. Boots and hooves thudded, some men sang, trumpeters sounded signals and history rumbled westwards. In the van rode armoured knights in their war splendour, pennants fluttering in the breeze, their helms slung on their saddle horns. It was a warm July day and the riverside breeze that riffled their silk caparisons was welcome. Behind the destriers marched men at arms in their heavy mail coats. Some with bowstrings strung taut and quivers quickly to hand, a company of Welsh archers made their way over the fraying surface of the old highway. And guarded by household knights, their squires on their ponies awaiting messages to be carried along the line, the long pennant with the three lions of England, yellow against red silk, was carried before the king.

Edward I had come north to humble the Scots, and from Berwick he led almost seven thousand men westwards. He knew his army was being watched, and although they rode and marched in full armour and mail in case of attack, show mattered too. It was a sight meant to hollow bellies.

Edward moved through a hostile countryside. Small groups of Scottish archers – some from the hunters of the Ettrick Forest – and crossbowmen could get close if the terrain allowed. Behind the thudding tramp of soldiers, the baggage train trundled, and behind that rode the rearguard. The destriers needed a supply of replacement shoes and hundreds were carried in stout carts, as well as the anvils for the farriers and charcoal for their forges. These horses could be fearsome in the ruck of battle. They were trained to bite and kick out with their metal-shod hooves. Big clench-nails could tear at skin, as well as deliver stunning blows. Sacks of corn to feed these horses and the soldiers followed, and behind them trotted movable supplies of food, herds of cattle and flocks of sheep. Edward's army was far too large to depend on foraging.

On 24 July 1301, the knights reined their destriers up the Long Track. Edward I had come to Selkirk Castle and almost certainly followed the Roman road that led on to Oakwood Fort. Almost 1,200 years earlier an army of about the same size had laid it, digging the drainage ditches on either side that can still be seen up at Hartwoodmyres. English scouts had swept the surrounding ridges before leading this huge medieval host through our farm. That summer, the rebuilding of the castle began and it may be that the king's pavilion was set up close to the mound and its ditches. Almost seven thousand men and all of their gear, supplies and camp followers needed a great deal of open space, a water source and security from Scottish guerrilla fighters who might have stolen upon them at night.

The farm was a good place for an army camp and it is highly likely that hundreds of pavilions, tents and shelters were erected on either side of the Long Track. With the Hartwood Loch to the south and the Haining Loch to the east, there was some protection from intruders, and the ridge of the Top Wood offered commanding views up the Ettrick and Yarrow Valleys, the direction from which trouble might come. The herds and flocks will have been safely corralled inside the pale of the Deer Park, and at the foot of its southern flank runs a small burn that feeds the Haining Loch. Its water is clear and sweet and our Old Boys in the East Meadow drink from it. When night fell on 25 July 1301, in the fields around our farmhouse hundreds of cooking fires will have flickered in the gloaming.

On that day, the Exchequer Rolls record that the king's paymasters doled out hundreds of thousands of silver pennies to the soldiers. Some of them were German mercenaries and one might have dropped a coin minted in the name of the Emperor Otto IV. Soldiers enjoy gambling and, with silver in their pouches, some will have played at the dice tables. In the half darkness, winnings might have been dropped and not picked up again until a metal detector buzzed in the twenty-first century.

It must have been a spectacular sight – and sound. English, Scots, Welsh, French, German and Gaelic rang out as men shouted for their horses to be saddled, their tents to be pulled down and food to be found. But it was not the first time – or the last – that an army would rumble up the Long Track.

16 April

Another turning moment in the year saw me take the log and kindling baskets up to the Wood Barn, where they are stored over the summer. Encouraged by the forecast of rising temperatures for the weekend, we hope not to see firewood in the farmhouse until late October. Meantime, many logs need to be cut from the piles of seasoned timber next to the barn.

I drove down to Kelso to find my grannie's grave. Armed with a map of the cemetery's zones, all identified by a letter, and a list of the headstones that stand either side of her unmarked lair, I was sure I would find her this time. But despite walking up and down the rows of headstones I could not. It was more than vexing, as though the earthly remains of Bina, her mother Annie and her aunt Bella had evaporated. Unmarked graves for unremarked lives. Not if I can help it. I shall ask for more information and go back until one day I find her.

17 April

At least some of Edward I's scouts and generals were aware that they had been advancing in the footsteps of history. For many centuries, Roman roads were known as strete, or streets, as in Watling Street and Ermine Street. It denoted a paved road rather than a mere cart track, something not built between *c.* AD 400 and the coming of the first metalled roads in the later nineteenth and early twentieth centuries.

18 April

A very misty morning before what might be a warm and sunny spring day. Dense mist and fog shrink the world, muffle sound and act as a continuing reveal, lifting a rolling curtain on a landscape that seems less familiar.

The fields of the East Meadow begin to emerge and the Nameless Burn trickles into the Haining Loch. Just beyond the edge of the mist, the tents and pavilions of Edward I's army are waking as grooms fit halters on the snorting destriers and take them down to the burn to drink, as men emerge into the chill of the early morning, stretching, shivering, looking around for somewhere secluded to relieve themselves. A smoky sun rises over Greenhill Heights.

19 April

As Maidie and I passed my son's house, Grace banged on the window. It was only a little after 7 a.m. but she was up, had breakfasted and wanted to come out into the misty morning to walk with me and my dog. It was a first, and it made the trip down the Long Track very different. From a perspective about three feet lower and sixty-five years younger, Grace sees things differently. The silver orb of the sun was showing through the mist and I said that it was unusual to be able to look directly at it. 'Yeah,' Grace mused. 'Weird.' When she came to the many lambs by the fence, the little girl did not send them scattering for their mothers as she lifted her hand and said, 'Hi guys.' I suppose she is too small to worry them. I watch all this and feel the tears come. Listening to Grace make her own story of this place moves me very much, probably because these are the early chapters of a long continuity. I hope so.

When we reached the dried-up puddles on the Long Track, we sang our mud song anyway. It is the first few lines of Flanders and Swann's 'Hippopotamus Song', with its chorus of 'Mud, mud, glorious mud, nothing quite like it for cooling the blood'. It is very surprising that this witty ditty should surface from an age that was bygone when I was young. Michael Flanders and Donald Swann were two dinner-jacketed performers who appeared on early black-and-white TV. Swann played the piano, while Flanders sat next to him in a wheelchair (he had contracted polio in 1946). They seemed to come from the world of concert parties and theatrical revue, the land of light entertainment. But the mud song is memorable and Grace likes it, especially the ending. When we get to the last line of the chorus – 'and there we will wallow in glorious MUD!' – the final word is an exultant shout.

Although it has been cold, the days of sun are working their transformative magic. Some of the red birches are in leaf, blossom is showing, just, in several thorns and geans, and the buds of the magnificent acer by the gates into the stable yard have begun to unfurl. The contrast of the colours – lime-green against a rich, deep magnolia – is unlikely but glorious. Textile manufacturers and fashion designers have long imitated the colour combinations they see in nature, but, so far as I am aware, I have never seen magnolia and lime-green used together. It should be.

From last year's diary, I see that the swallows came yesterday. It has been too cold for them, and the flies they need, and the winds have come from the east, the wrong direction. But the weekend forecast is good and the wind will be back to the south-west, the right direction.

This is Good Friday, the first of the three days of Easter, a holiday when many take a spring break and much chocolate is eaten. It was the day Christ was crucified, a cruel and lingering death used by the Romans to advertise the conse-quences of breaking the law. Swift executions like beheading were over too quickly and many more would see the agony of the cross and be intimidated by it. It is a gruesome image to celebrate, but of course the power of Easter also comes from the Resurrection two days later. The principal Christian festival from which others follow, the date of Easter was settled at the Synod of Whitby in 664 after a long controversy. The Celtic church based on Columba's mission and Iona had been very influential in Northumbria, founding the monastery of Old Melrose only a few miles from the farm. Its bishops reck-oned the date differently and all sorts of political pressures pushed for uniformity. But one that is frequently forgotten relates to a battle no longer fought.

At Easter, the war with Satan and evil was at its height and to defeat the forces of darkness, or at least keep them at bay for another year, God needed to be supported by all who believed in Him. And the way in which Christians could do that was by praying. Like a battle, numbers mattered, and if some Christians were praying at a different time, because they reckoned the date of Easter fell on another day, then God's army was much reduced. All Christians needed to pray for the defeat of Satan at the same time.

20 April

By calling it a bank holiday weekend, many people seem to prefer to relate Easter to money rather than Christ's Passion, his death and resurrection. It is a strange and surprising switch. This morning's tabloid headlines warned of sweltering temperatures in the eighties, and so when I put on a suit to drive down to Kelso it felt like far too many clothes. But I was determined not to turn up in a shirt and shorts because that would have been disrespectful. With the kindness and help of the cemeteries department of the Scottish Borders Council, my grannie's grave had at last been found. I wanted to go and see it for the first time, to attend the funeral I had missed all those years ago.

Only moments after I walked through the gates of the old cemetery, I saw the forlorn little white peg placed by the foreman. In a corner, shaded by a huge holly tree and a newly trimmed yew bush, he had written on it: Moffat G187. I took off my hat and after a while hunkered down on the ground where my gran had been buried in 1971 when I had been in Italy. Bina was laid down over her mother Annie, who had died in 1936, and before that, in 1928, her aunt Bella had been lowered into the lair. It felt as though I had walked around a wide circle to come at last to this shaded, forgotten corner.

I had been unable to find my grannie's grave because some of the lairs to her right also had no gravestones and hence no names I could match with the list I had been given. The one for Jessie Oliver that I might have recognised had collapsed. Old bricks and a lump of broken concrete lay to one side and over the years the yew had begun to encroach. It seemed that she and her mother and aunt were being squeezed out of history. Yew trees are often found in churchyards and cemeteries because they live to a great age and remember a pre-Christian sanctity, and I consoled myself with the ancient notion that when the roots reach down to the dead, they touch them and release their souls through the tree and on into the sky above. Next to Bina's lair, I could see roots as thick as a forearm showing through the grass.

Determined not to leave my gran in this dusty corner, I walked down to the Tweed to a garden centre to buy a flower holder that could be set in front of a tombstone, even though Bina did not have one. A gormless girl had no idea what I was talking about, but an older lady advised me to try the florist in the town. She turned out to be the soul of kindness and, with two bunches of yellow carnations in a conical holder filled with water, I walked back to the cemetery to bring some spring colour to my gran's grave. Carnations would last, said the florist.

I don't know how these matters are arranged, but I wondered if I should be buried with Bina so that she could keep me close in death as she had cuddled me in life. I stood for a long time, thinking what I should say, out loud. Love is by no means a given in families, but my gran gave me unconditional love and I wanted to thank her for that, no matter how banal it might have sounded. And then for a few moments I wept for the loss, for the passing of almost fifty years since I had seen her. But at last she was found and I promised to come back with more flowers. And with my sisters' help, she and her mother and aunt will have a head-stone inscribed with their names and the span of their lives instead of a little white peg.

21 April

To lift my sadness, last night the swallows came. Six swooped over the stable yard, diving between the boughs of the old oak and the plantation of Norway spruce and larch behind the hay barn. Having completed their astonishing journey from Africa, these little daredevils had come back home and they celebrated with some virtuoso aerobatics, synchronised swirls and turns, doubling back on themselves and flying at breakneck speed into the stable boxes where the old nests are. Then they disappeared, off over the fields for an evening feed. It was a wonderful moment.

Grace has taken to coming out with Maidie and me on our early morning walk. Looking very grown-up in a denim jacket, pink leggings and blue wellies, she marched down the Long Track to see the lambs. On the way home, she looked up at me and said something that turned my heart. 'I had a wee sister once.' And she held out her hands and arms as though she was cradling a baby. I thought Grace was talking about her older sister, Hannah, who died stillborn. But when I did not, could not, say anything the wee lass carried on, explaining that babies came out of tummies and when she was a flower girl at Auntie Beth's wedding, that was when she had the wee baby. It was a three year old's story, woven from overheard fragments, nothing to do with Hannah, but it brought more tears. I wiped them away before my granddaughter noticed and we walked on down to the farmhouse.

Wherever her parents decide Hannah's ashes should be buried, I shall be buried there beside her, instead of with my gran. I won't let the little one be alone for too long.

22 April

Up at Windy Gates with Maidie early this morning, I heard the clank of the jaws of the harvester in the distance, a dinosaur-like machine with a swinging saw-arm that can cut down tall trees in ten seconds. When we climbed up a little at Huppanova, I could see that the wood had now been clear-felled to reveal . . . a mobile phone mast. But the shape of the land, hidden for so long by the sitkas, was laid bare. A few hardwood trees were left as lonely survivors amidst the carnage. Soon the summer's growth will begin to hide it and let the land heal a little.

It was a sight that reminded me of how much the land can change – and quickly. When Edward I's army camped around our farm, there were no fences, no dense plantations and indeed very many fewer trees. All the colours of the landscape were natural, the hundred shades of green, brown, grey and yellow. And the silence was only punctuated by the calls of birds or the lowing and bleating of animals. Against the deep camouflage of the land, farmhouses could often only be made out by the spiral of smoke from their cooking fires. When the royal and noble silk standards fluttered and the destriers were clothed in their caparisons covered with heraldry, vivid colour will have splashed, something rarely seen. And the landscape will have rung with the call of orders and the clatter of movement before the army rumbled on.

In the early fourteenth century, after Robert Bruce defeated Edward II at Bannockburn in 1314, the climate began to change the look of the land radically. When the triumphant Bruce surrounded Carlisle in the summer of 1315, it rained so torrentially that his army and its siege engines became irretrievably bogged down in bottomless mud around the city walls. After a short time, all efforts to capture Carlisle were abandoned and a sodden Scottish army moved on. That autumn waterlogged fields produced a meagre harvest and the following year was little better. Hartwood Loch will have refilled and looked much like it did to the flint-knappers who lived on its margins five thousand years before. The Little Ice Age was beginning, a period of about five hundred years that saw long and regular periods of bad weather. Famine frequently stalked the land and it was not until the 1850s that there was sustained improvement. Now we face more extreme fluctuations. When the sun shines as bright as it has for the past few days, my enjoyment of it is heavily qualified by anxiety about more climate change, this time brought on by man-made causes that will be almost impossible to halt unless action is taken, not soon but now.

23 April

Last evening I met Walter Elliot and Rory Low, the metal detectorist who had found the medieval coins around the farm. Washed and polished after centuries in the soil, Walter called these small, glinting objects treasure, but not for their value as bullion. In reality they are thin and very light. Their value is in the knowledge they give of the past. The German coin with the head of the Emperor Otto IV was fascinating. Looked at through a magnifying glass, it was possible to see that his crown followed a Byzantine model, with side-pieces like large pendant jewels. The wafer-thin silver penny of Robert II of Scotland looked as if it had been newly minted and the image of this, the first Stewart king, was very clear and very regal. Sadly the historical reality was very different. Having succeeded to the throne in 1371 at the advanced age of fifty-five, Robert turned out to be so ineffectual that the royal council removed his power to govern in 1384 and appointed a regent. Such was his miserable, beleaguered demeanour, the king was nicknamed Auld Bleary.

After Rory had packed away his coins and the other items he had found in the fields (many legs for iron and bronze pots used for cooking but, strangely, very little metal horse gear), we drove up to the top of the Deer Park. I realised I had not been up to the plateau for at least a year. Walter wanted to look at the probable site of the Ogham stone I had found because he thought it might have been a sacred enclosure. Perhaps there were graves, perhaps facing east to west, something characteristic of Christian burial. But there was very little reaction to his divining rods and, with his detector, Rory found only some spent cartridges from the Second World War.

Nevertheless, it was good to be up at the top of the park. The views to the west and the dying sun were heart-filling and we stood in that intensely atmospheric place and talked about nothing very much for a time before descending. On the way down, I noticed that a majestic, mature beech tree had been blown down in the summer storms of last year, its leaves having acted like fatal sails in the seventy-mile-an-hour winds. They were still on its boughs, rust-coloured and brittle. The roots had been torn clean out of the ground and the thick trunk was exactly horizontal. I was vexed to see that and will ask a chainsaw operator to log it for me. It is more than I can manage with my saw. My brother-in-law is a talented woodworker and he might use some of this beautiful, pale-cream-coloured wood to make a table or a stool. That way it might live on.

In this morning's early and warm sun I watched my son and Grace take the two big dogs up the track. The wee lass skipped for a few steps and her walk was jaunty. Why do little children skip? I think it must be joy that prompts them, the joy of being alive and with lives stretching out before them on a beautiful spring morning.

24 April

Where the Bottom Track turns through ninety degrees to become the Top Track, someone who used to help around the farm parked several saplings in the banking. He once worked as a forester and sometimes did that to help these sticks root before moving them elsewhere. Except neither he nor we ever got around to doing that. Now, about five or six years later, in a space less than a square metre, there are three American red oaks, two

Norway spruces, two trees I could not identify and a larch. The larch has climbed highest but, because of their position, all are well watered and in sun for most of the day.

Just beyond the fences by the side of the Long Track, I saw the first and so far only casualty of this year's lambing lying in the grass, its eyes pecked out by crows. Spring growth followed by spring death.

27 April

It was a damp, muted morning after overnight rain that will help the grass as temperatures rise. One of the mares, Wendy, is carrying a foal that will drop at the end of May and good grass at this time in her pregnancy should help. I plan to plant out tomatoes this afternoon on the terrace in the hope that they will thrive outside this summer. In the past I would have worried about rabbits eating the sweet shoots but several times in the last few days I have seen a stoat prowling the terrace and the garden. Not only will he or she keep the mice and voles manageable, they will also keep rabbits at bay. Spring is when their kits are born and it seemed to me that our stoat was out hunting to feed his mate who would be suckling their young wherever their nest is. This seems to me to be a tomato-friendly micro-ecosystem working well.

28 April

Lichen grows everywhere on our farm: on tree bark, on the old stones of dykes, even creeping over an abandoned field gate in the long grass of the Top Wood. It is delicate, beautiful, very rewarding close up. The deeply corrugated bark of older elder trees is often covered with a patchwork of a yellow lichen of great subtlety and density even on cloudy mornings. Sometimes rich egg-yolk, sometimes the pale yellow of the waning moon.

Clustered in small filigree clumps on the branches of the birches on the fringes of our woods, light grey lichen grips so tight that not even last year's summer storms can shift it.

When the farmhouse was rebuilt in 1994, the roof was lifted up to allow a second storey of bigger and higher bedrooms. The steeply angled gables were formed by precisely cut slabs of sandstone copes that rested on corbelling projecting out from the walls. For years they were the colour of freshly baked shortbread and looked very new compared to the grey whinstone of the old house, but now they are patterned with patches of green and grey lichen and spotted with a white variety that looks like accidental splodges of dripping paint.

It turns out that these crusty, bushy and sometimes shrub-like plants are not plants. Even though they seem very adhesive on trees and stones, they have no roots and absorb water not from the surfaces they colonise but through photosynthesis in the air. And they grow in profusion where the air is pure, although some of the thousands of varieties can tolerate pollution. Apparently lichen can also survive in space. A Russian crew released a canister with lichen in it and exposed it to the airless atmosphere for fifteen days before bringing it back to Earth unaffected. A little more research told me that a staggering 7 per cent of the planet's surface is covered by lichen and some varieties can live for ten thousand years. I found that a comfort.

All I knew about lichen before we came here was that in the Hebrides weavers harvested it to make dye. The muted, subtle colours of Harris Tweed come from what the Gaels call *crotal*. It can also be used as a simile that made me smile when I first heard it. A friend of mine who lived on Skye went bald but retained a patch of hair above his forehead and his alleged friends called him Donnie Crotal because his scalp looked like patches of lichen on smooth rocks. Here, nature has painted our farm and its trees and stones in a hundred colours, many of which will still flourish long after we are gone.

29 April

Rory Low, the metal detectorist, has found something strange, unexpected. On its way to Selkirk Castle, the Long Track runs spear-straight through the Doocot Field. Dovecot in English, it is so-called for the ruin of a large stone building whose southern wall has collapsed to reveal about a hundred nesting boxes. As gardeners fed them with grain and other morsels, doves or pigeons were encouraged to multiply there and supply a handy source of meat for the occupants of the big Georgian house at the Haining.

Rory has found a lot of English and some Scottish silver pennies in a wide cluster around the doocot, strongly suggesting that this was the centre of the English army camp in the summer of 1301. But yesterday he came across something else: a very thick piece of heavily patinated lead. Found deep, twelve inches down, in the same stratum as the coins, it seems to be a large fragment of roofing material. It has grooves along one edge that suggest it was formed over some other material, stone, wood, perhaps glass. The rich patina and the thickness of the lead suggest it is very old, perhaps medieval, like the dated coins. Was there once a building in the Doocot Field, by the Long Track? A chapel?

We plan another field walk with Walter and his divining rods, and I will scour old estate maps I have acquired.

30 April

Bickering like an old married couple about which way to go, two Canada geese flew round and round the Tile Field before heading east towards the sun, the agitated honking fading as they breasted Greenhill Heights.

May

1 May

On a still May morning of sticky buds, lustrous leaves and a soft rain, I was thinking not of new life but of death. For the Scottish Borders, the fourteenth century was little more than a mournful procession of disasters. When Robert Bruce died, he was succeeded by David II, a child only five years old who turned out to be a hapless king. After a costly defeat at Halidon Hill near Berwick-upon-Tweed and the loss of the town in 1333, Scotland was riven by civil war as Edward Baliol tried to wrest back the crown from the Bruce line. His father, King John Baliol, had been summarily deposed by Edward I in 1296, and at Stracathro in Angus he was humiliated as the royal arms were ripped from his tabard. When Edward I rode south, he is said to have remarked that 'a man does good business when he rids himself of a turd'.

Baliol's son styled himself Edward I of Scotland and it suited Edward III to support him. Armies marched back and forth across the Tweed Valley, trailing tremendous destruction in their wake. But there was much worse to come.

In June 1348, the Black Death made landfall in England, having raged across Europe. A year later, the pandemic had yet to reach Scotland. The more simple-minded called it 'the foul English pestilence', believing that the Scots were naturally immune because they were Scots. They also believed that Englishmen had tails. About six miles from our farm, at

Caddonlea, an army massed to invade England and take advantage of the chaos wreaked by the plague. Probably brought by mercenaries, the fatal disease suddenly erupted amongst the campfires and the army of opportunists panicked and scattered. The south-east of Scotland suffered badly after 1349, with about one-third of the population struck down. Fields were left uncultivated, beasts untended and the local economy plummeted into a long recession. And none of this was helped by the weather. In the second half of the fourteenth century famine struck twice and recovery was slow. An undernourished population was very vulnerable to disease and the plague returned in 1361 and 1369, with mortality rates of between 10 per cent and 15 per cent.

The past is fascinating, but sometimes not a time I would want to return to.

2 May

Across the battle-strewn landscape, its population fearful of repeated returns of the great pestilence, there were green shoots of hope. Difficult to date but easier to detect, these took the form of appropriation, as ordinary people began to assert their rights. Selkirk's royal charter was probably granted sometime in the twelfth century when the castle was built and the Deer Park established, and as a consequence of some lost legal grant, or perhaps through custom and practice, the townspeople began to use the land to the north and south. There they grazed their cattle and sheep, dug peat for their fires, gathered kindling and cut bracken to cover their floors and thatch their roofs.

Likely land originally in the king's possession – what became Selkirk Common – included our little farm. The king and his town were the first recorded owners of the fields and tracks we have come to know so well. We know that the farm formed part of the South Common because its boundary, the Common

Burn, runs behind the stable yard. It is an ancient margin and on each side its banks have been built up higher, not to minimise flood risk but to make its course more emphatic in the landscape. Beyond its banks lay Hartwoodburn Farm, just as it does now, and it was privately owned – the resort of villains. Sometime before a circuit court convened in Selkirk in the summer of 1510, the farmer at Hartwoodburn and his men led his beasts across the burn so that they could graze our fields, even though at that time they were owned by the burgesses of Selkirk. In court, the sheriff found the Hartwoodburn men guilty and probably forced them to pay a fine.

The burgesses were zealous in protecting their common against encroachment by surrounding lairds and, in 1541, John Muthag, an early provost of Selkirk, and his baillie, James Keyne, went to Edinburgh to plead a case at the Court of Session. They were ambushed and murdered somewhere on the road. The town could not survive without its common, and it was indeed a matter of life and death. Until recently, food and fuel came directly from the lands to the north and south of Selkirk. Most townspeople kept cows for milk, cheese and butter, sheep for wool, milk, cheese and meat, and they still gathered the seasonal wild harvest of fruits and berries, wood and kindling. To the north rises the low hump of Peat Law and on its flanks the remains of the banks where peat was dug and dried are still visible. Now, almost all of Selkirk's food and fuel arrives at its shops in articulated lorries from large depots, often a long way away, and an ancient bond with the land has been broken.

In a hill town in southern Tuscany this umbilical connection still survives and its history has recently been commemorated by an evocative and beautifully formed bronze sculpture in the main piazza. I know Pitigliano very well and used to stay there, sometimes to work. In 2008 a dignified and attractive ceremony unveiled *Il Villano e il suo asino*, the countryman (or peasant) with his donkey. He is putting a wooden pack saddle on the little donkey (whose head is down, looking a little weary) before going out to work in the fields around the town. His hat and clothes are post-war in style, and until the end of the 1950s townsmen led their donkeys under the medieval archway every morning. Their fields are still there, a pattern of lush cultivation, rows of vines, tomato plants, zucchini beds, dreels of potatoes. All of that glorious profusion, well-watered and glowing in the sunshine, can be seen at the foot of the cliffs the town perches on. Now, the donkeys have been replaced with little three-wheeled trucks known as Piaggio Ape that sound like bronchitic lawnmowers.

In the narrow and shaded medieval streets of Pitigliano, houses have what are known as *cantine*, cellars cut into the soft volcanic rock called *tufa*. These long tunnels often burrow underground for several metres. Very cool, even in the midday Tuscan sun, some were used to stable the donkeys and many as stores for the food grown in the fields. Hams hang from hooks and the excellent local white wine matures in huge barrels called *botte*. I remember seeing an old lady going into her *cantina* with two empty two-litre plastic bottles and coming out with both filled to the neck with the characteristic greenish wine. It had a white, frothy head. I wished her good morning and, as we chatted, she put the bottles back in the shade of the old oak doors of the *cantina* to keep them cool. Shelves were lined with rows of large jars full of preserved fruit and vegetables, and several rounds of pecorino cheese were stored well beyond the reach of mice, waiting for winter. There were apples set on what looked like brown-papered shelves and a basket of large tomatoes that looked like pumpkins. The summer harvest around Pitigliano is long, lasting from June to early October, and its fruits keep well in the cool *cantine*.

Small medieval Scottish towns worked in the same way. Early in the morning donkeys and sometimes carts were trundled out of Selkirk's West Port and onto the fields of the common around our farm. Buckets were brought to milk the cows and probably carried back suspended from milkmaids' yokes. If each bucket was a similar size and filled with roughly the same volume, these yokes not only helped with the heavy weight, they were remarkably stable, rarely spilling a drop. Other sorts of food, such as barley and oats, were harvested and stored in the townspeople's cellars as long as possible, sealed tight to keep out vermin. Winter forage was also essential to keep a cow fed. And so when the Hartwoodburn men came up before the sheriff and were convicted of illegally grazing our fields, I hope their fines were hefty.

3 May

This morning I was much moved by a trivial, simple detail. Amongst the birches on the margins of the Bottom Track, one with a rich, red bark took a terrible beating in last year's summer storms. About halfway up its peeling trunk, a major limb had almost been torn off and come to rest not on the ground but on the branches of a sitka spruce that stands beside it. To my great surprise, the bough had kept enough of a connection with the trunk to come into leaf this spring and help the birch to photosynthesise and grow. I liked that – mutual support amongst the community of the trees.

4 May

It is very cold this morning, a bitter wind blowing out of the north-east that feels as though it has come straight off the wastes of Siberia. Even though the sun is brilliant in a cloudless, cerulean blue sky, the eye-blearing, ear-nipping cold means it does not feel like a May morning. The buds are closed tight against the overnight frost and, having been out all night in it, the horses in the Tile Field stand motionless, side on to the rising sun, soaking up its warmth across as much of their body area as possible. Maidie and I hurry on down the Long Track.

I read yesterday that by 2050 two-thirds of the world's population will live in large cities. Even now there are hundreds in China and India that are so new I don't recognise their names. These migrations mean that a mass consciousness of the natural rhythms of the planet will be much reduced and most people will cease to observe the cycle of growth through the seasons. There are satellite photos taken in darkness that show dense concentrations of light in Western Europe, the USA, India and China. That means many fewer people can see the stars and planets of the night sky. Far from being cosmopolitan, cities close down the world and the cosmos, shrink horizons and diminish our sense of how the world's seasons change.

On our farm, I have learned that the weather and not the calendar turns time. At the beginning of May last year, a long, warm and dry period began with average temperatures double this morning's values at over seventy degrees Fahrenheit (about twenty-one degrees Celsius). These conditions created a different landscape. A year ago I could smell the sweet scent of the stand of poplars at the gate into the Deer Park and the strawberries-and-cream blossom of the geans was everywhere. In the raised beds, my early potatoes had pushed up into the light (only to be earthed over) and the plum and apple trees were white with the flowers that would become fruit. The well-worn and much misunderstood phrase advising caution, 'Ne'er cast a clout till May is out', has nothing to do with the calendar. Winter clothes should only go back in the wardrobe when the blossom on the May Tree is out, a sign that the year has turned and the weather warmed.

5 May

Potatoes are much cheaper to buy than to grow when the cost of labour is taken into account, but last year's earlies tasted sweet and earthy at the same time, a subtlety not available in the shops. And when they were cut prior to cooking, the juice shot out as though they were lemons. Even so, their clean, fresh taste is not why I grow the humble potato. I want to be involved in the cycle, have an annual harvest of some small sort and feel that our land could be productive. As the years roll on, I hope to plant more vegetables, rotate crops and build more raised beds. In the windless warmth of the old conservatory, the tomatoes are a riot of burgeoning greenery that will soon need support. Like me.

6 May

Illness prevented the planned survey of the area of the Doocot Field where Rory dug out the piece of lead that was suggestive of a structure of some sort. We hope to get out next weekend. But meanwhile the mystery has deepened. Using baby oil to clean the piece of lead and show up any detail, Rory has discovered a surprising series of markings along the thickest edge. A row of vertical and precisely horizontal straight lines cut with a chisel, a burin or some other sharp tool, they are not accidental. But what do they mean?

At first Rory thought they might be a line of Ogham, the tree language I found on the inscribed stone in the Deer Park, but they don't look right to me. Ogham is usually arranged on either side of an edge, an analogy for the trunk of a tree, and the stones sometimes have two flat planes. Or the inscriptions can be vertical, often cut into stones carved for other uses, such as standing stones, tombstones or crosses. Many that were carved on flat planes have a clear stemline that the letters cross or are attached to. These marks on the lead are not arranged in any of those patterns. I wonder if they are tally marks, a sequence of early arithmetic. In an age before paper, workmen made notes on all sorts of surfaces, including stone and lead. The marks look a little like five-bar gates, a common method of tallying I still sometimes use.

Which leads to another question. What was being counted? Quantities in a building project? And if so, where was the building? It could have stood some distance from the find-spot. As it is now, lead was valuable and Rory might have retrieved a chunk of stolen property.

Enigmas like this are exhilarating and they prompt the past to come racing back across the centuries, make a grass field nibbled by ewes and their lambs come alive, reminding us that we are only one of many, the most recent generation to walk our lives under the big skies of the Borders. The dozens of silver pennies Rory has discovered remember the workings of a money economy based on precious metals, carried in pouches and lost on a forgotten journey up the Long Track. They are resonant echoes of transactions we can understand, and sometimes they are very surprising, like the imperial coin of Otto IV from Germany. But the piece of lead is different, perhaps impossible to understand. Is it evidence of a crime, the coldest of cold cases, or a tantalising fragment of an important building long lost in the grass? The wooden houses of our medieval ancestors were almost entirely built from organic, perishable materials. They did not use lead to keep their roofs watertight or their windows draught-proof. Somewhere important, a building of high status, is hiding from us and so is its story.

My morning walk with Maidie through a soft drizzle revealed an image of cheer, of hope. In a moss-covered old tree trunk by the edges of the Bottom Track that I had thought was a boulder, two birch saplings have seeded, their leaves bright lime-green and flushed with vigour. Inside that old stump lies a store of ancient goodness. And the sight of such unlikely fecundity sparked another thought. The underground web of thread-like connections between trees – old, young and even apparently dead – that is so memorably described in Peter Wohlleben's magical book *The Hidden Life of Trees* may suggest an alternative to Darwinian evolution. Perhaps there is co-operation as well as competition in the natural world. The fittest trees survive with the help of their ancestors.

10 May

After three days away for work, when I took Maidie out this morning she greeted me as though she had seen me yesterday and not last Monday. I imagine that is how canine memory works; habits of life trump the ticking of the calendar. As we walked up the Bottom Track, it was strewn with blossom snow. The small, white flowers of the geans had been falling in the overnight breeze. On the Top Track there were welcome puddles in the potholes from two rainy days when I had been in Ireland. It surprised me how quickly I could switch from thinking about making television to the concerns of our farm.

Up in the Deer Park, the western vistas were lit by a cold sun and I began to make plans for the summer, listing what needed to be done. Potatoes should be earthed over to keep the tender leaves from overnight frosts, tomatoes need to be trained, lettuces planted in the old conservatory and a winter store of logs cut. The circle turns constantly.

12 May

On this windless morning the land lay still under an open sky, the sun brilliant and warming. The buds of the big gean at Windy Gates are slowly unclenching and even the spindly birch near it is putting out shoots and leaves. It has always been a sickly little tree, its roots probably wrapped around a big stone. While it has reached only half the height of the avenue of trees by the side of the Top Track, I sympathise with its gameness as it struggles for life and I like the yellow lichen on its lower limbs. It looks to me like the little kid in the corner of the playground, hoping to evade the notice of the big bruisers making all that noise.

We have many fewer swallows this spring and so far no house martins have come north to us from Africa. I fear that the crash of insect populations is beginning to shudder through the food chain and climate change may be claiming more early casualties before its effects really begin to bite. But the sparrows, once under threat, have come back and they have colonised the martins' old nests above the porch. Plump, pleased with themselves, bold and bossy, I like their attitude.

13 May

Up early, Grace banged on the window as Maidie and I passed, on our way up the Bottom Track on another sun-flushed morning. In a red Minnie Mouse dress, warm tights, blue wellies, a quilted jacket with important pockets and an orange baseball cap, the three-year-old splashed colour in the green landscape and I began to see it through her eyes. First we watched the wee Westie sniff the tracks of the naughty rabbits as they wiggled between the trees into the deep, dark wood. Where there are pixies . . . On the Top Track we sang a nonsense song and at Windy Gates turned to look for the Dangerous Brothers, the lambs careering around the fields at Huppanova.

Maidie did a poop and was given a treat by Grace. And then we walked off together into the sunlit Deer Park. Grace took my middle finger in her hand, still too small to grasp any more of mine, and I felt tears prickle. I'm not sure why. These moments are precious, I know, and I realise that I will probably not live to see the little one fully flower into a sparkling young woman. But I often think that my control of my emotions, never tight at the best of times, is slackening, close to embarrassing.

14 May

To those who lived through the wet, frost and snow of the Little Ice Age and who survived the Black Death of the fourteenth century, it must have seemed that God was often angry. Plague returned to the Borders in the early sixteenth century and a sustained period of poor summers produced meagre harvests. Hunger sharpened attitudes in ways we no long experience. Food was eked out and nothing thrown away. My grannie Bina remembered a life when there were occasional shortages, if harvests had been disappointing, and she was horrified if my sisters and I did not clean our plates. A little chubbiness around the middle was seen as good insurance.

'Kitchen your meat!' was her motto, and by that she meant that any morsel of ham or beef should be well padded out with various sorts of carbohydrates. On a round, thick, black, cast-iron griddle (which she pronounced 'girdle', an example of metathesis, the shifting of consonants, common amongst country people – Bina talked of pattrens and new things were modren) that she set directly on the highest flame the old, grey gas cooker could muster, all sorts of old-fashioned, filling foods were made.

After mixing a paste of oatmeal, flour, fat and some water, she rolled out thick oatcakes and laid them on the griddle to sizzle and pop. A much more runny dough was poured on in dollops to make sweet dropscones and she also made buttery, crunchy, hard biscuits. Into the oven went scones and bannocks. Very little bread was eaten, and Bina told me that in the cottage at Cliftonhill where she was born and raised it was a treat to have what she called 'wheaten' bread. Shop-bought and white, it was eaten on special occasions with butter and jam. Dumplings called doughboys were dropped into stews to bulk them out and in Bina's kitchen absolutely nothing was wasted. Anything she could not immediately think of a use for was tipped into the ever-simmering soup pan.

15 May

Blearily making tea in the kitchen at about 6 a.m. after a restless, tossing, sultry night, a flash of rapid flight flickered on the edge of my field of vision and made me turn my head quickly. And suddenly they were here. Four house martins, unmistakable with the distinctive white bands across their tails, were flying in and out of the corner of the terrace where they built last year's nests. I think they had just arrived, found their way back home after a six thousand-mile journey from Africa, crossing deserts, seas, mountains and up the length of Britain. Sparrows have been investigating the old pods of mud that clung to the eaves through the winter and so these four weary travellers might seek another site to settle and raise their broods. But they are here, wherever they nest, and I am much cheered.

Out early on another sun-blessed morning, I felt that warmth and light make the world a room, a place to linger, to sit down on a grassy bank, the dew long dried, and watch and listen. There was a peace on our place this morning, only birdsong to be heard, and all of the animals seemed content: the ewes moving slowly, nibbling the sweet spring grass and their lambs for once not on springs. We turned along the Deer Park track and past the badger setts. Mounds of excavated earth and several well-worn entrances and exits on the edge of the wood suggest an underground complex of tunnels and many chambers. I thought of the little grey bears curled up in the cool, full of last night's hunting and scavenging, sleeping sound while the world's room warmed above them.

Thinking that we were bringing food, Suzie and Rosie, our two old mares, came walking up an old track, past a bank of egg-yolk yellow gorse. The sun was behind them as they plodded unhurried towards us. As I rubbed their velvet noses, their breath smelled of fresh morning grass and, without their rugs in the heat, I could see that the good weather after last week's rain had helped them put on glossy condition.

At the foot of the western flanks of the Deer Park we buried Tom, my daughter Beth's old pony and the first horse we bought and learned from. Full of feisty character and talent, Tom was much loved and we planted a rowan for remembrance over his grave. Like he did, it exudes life and vigour.

Summer storms two years ago brought down two major limbs from a stately old chestnut tree near where I found the Ogham stone. Most of the fallen wood has been off the ground and, with the bark beginning to flake off, I could see that it was well seasoned. When I get round to organising a logging weekend with Adam, Beth and her husband Ross, we will harvest that dense wood, and all its years of growing will burn bright in the winter fires. Last spring I noticed a sapling growing where we dumped many tons of topsoil (after digging out an arena next to the stables) by the Deer Park track. Now I see it is a tall, leggy tree, a child of the chestnut with the fallen limbs. Its roots will reach down into the goodness of the topsoil and soon grow to an established maturity. I hope to see it.

18 May

After welcome overnight rain followed by a soft drizzle this morning, the land has come alive. Leaves have unfurled and are almost fully out, the warmth and the wet having worked their magic. The scent of fresh is everywhere. The last big tree near the farmhouse to come out is the ash I planted by our boundary fence, and its shoots are now showing. More showers are forecast for the next few days, and when the sun returns the hundred shades of green will once more colour the landscape.

19 May

The land hides its memories, keeps its secrets. The homely geometry of green fields, thorn hedges, fences, tracks and woodland is recent, no older than the two or three centuries since the enclosures and the rapid modernisation of farming. But just below the surface of a rural landscape many believe to be old and traditional, even timeless, there lies a faded map watermarked and stained with blood and history.

This morning Grace knocked on her window as Maidie and I passed on our way out for a morning walk. My granddaughter wanted to see the Dangerous Brothers, the lambs that career around the Doocot Field in endless wacky races, exulting in life just as Grace does. She is fuelled by toast and Coco Pops and they thrive on sweet spring grass. On the tracks we saw rabbit poop, big bird poop, sheep poop and fox poop I had to drag Maidie away from. 'There's poop everywhere!' exclaimed the wee lass, her palms upturned in cheerful, exaggerated exasperation. We rescued a drowning snail from one of the puddles on the Top Track. When we entered the Deep Dark Woods, we found a ground nest near the old fence. It was a rough circle of flattened grass and willowherb about two feet across. Something large, like a deer, had made it overnight, sheltering under the trees from the showers. Grace knelt down for a closer look and when I pointed out deer tracks in the mud nearby, she wanted to follow 'to see if it is OK'.

Grace loves the land, and our farm is her kingdom. Her story of it is fresh and constantly renewing just as mine is deepening. Rory emailed last night with photos of more finds from the fields we walked through this morning. Many small blobs of lead have come up and a new metal detector seems to be locating much more of the detritus of war. He thinks the blobs have been dropped on the ground during the process of metalworking, probably the production of lead shot. There is so much hidden deep in the field that it may have been a substantial operation. But why were musket and pistol balls of the seventeenth and eighteenth centuries being cast there, only a short distance from our farmhouse? There should be the outline of buildings of some kind, however temporary. Perhaps we will find out next weekend, when Walter Elliot brings his divining rods and his expert eye for the lie of the land.

Rory has also found fragments of twentieth-century ordnance. Brass base plates of Mills bombs come up regularly. These were the first hand grenades, used in the trenches of Flanders from 1915 onwards. The plate Rory gave me has WD on the inside, for War Department, and a capital Q for Qualcast, the manufacturer, much more famous for making lawnmowers. Two sorts of grenades were manufactured. The more familiar is the grenade that was thrown in the classic manner, overarm with the thrower immediately ducking down out of the blast. The edge of its range was about one hundred feet and cricketers were thought to make the best grenade throwers. Rifle grenades could be fired about one hundred and fifty yards but were even more dangerous to launch. A soldier had to seat the butt of his rifle in the ground because the recoil of firing such a heavy object was too great for his shoulder. When the pin was pulled, he had only five seconds to pull the trigger to fire a blank that would in turn fire the grenade. If the rifle jammed, there was nowhere to take cover. It seems that essential training for this particularly dangerous art of war took place in the fields of our farm and in the Deer Park.

Last night Rory sent me a photo of a shell primer he had dug up in the Doocot Field. It came from a 1941 British 25-pounder gun, a standard piece of mobile artillery in the army for twenty-five years, only decommissioned in the 1960s. Its design is instantly recognisable from many war films and TV documentaries. A two-wheeled gun with its barrel projecting from a wide, protective metal screen, it could be set up quickly and fire rapidly. Depending on the elevation of the barrel, its range was between five miles and six miles. Which begs a question: where was the shell that Rory found fired from? If these guns were trundled onto our farm for training exercises, as they must have been, where were they set up? My only answer is the high plateau of the Deer Park. Its long vistas out to the west will have allowed instructors to see where shells landed, whether or not they were on target. Perhaps they detonated. Not only was the detritus of war left lying, the projection of grenades and shells will have roared through the sky before exploding on this peaceful landscape where the Dangerous Brothers career around.

20 May

Out with Grace, I have to bear in mind when pointing out things that she is half my height and sees the world from three feet and not six. Maidie sees it from about ten inches, fourteen if she stands up straight, a very different perspective. Sometimes I think her lack of height frustrates her and she often scrambles onto a tree stump for a longer view. To her, almost everything is close-up. At fourteen inches, the sides of the Long Track are high and it must seem like a canyon. Every patch of ground she crosses is examined with great intensity, her head swinging from side to side as she sniffs after the scent of animals who passed that way before she did.

Maidie has a jaunty gait – her little paws seem to bounce off the track, and when she is navigating her way around a particularly interesting sniff she often lifts a front leg to avoid too much disturbance. I have seen her hop daintily in a confined space, turning on a sixpence. Outdoors she sees everything from low down, but in the house she leaps up to the backs of chairs and sofas to look out of the many windows. Searching distant horizons for danger, her tail is an indicator of any threat level, and there are many. Straight up is DEFCON 1, all the way down is zero. Her sudden barking at God knows what annoys me (I think she finds some trees and bushes irritating) and I often shout at her. But all she is doing is protecting the pack of which she is a (crucial) member.

Never having had one before, it has taken time for me to get used to the terrier temperament. They bark a great deal and are naturally aggressive. Even though she is younger and smaller than our labradoodles, Lillie and Freydo, Maidie has asserted herself as Top Dog and there have been one or two scraps that have gone beyond playful. She is certainly not a lapdog, but she does follow me around, and when we sit in the porch in the evening with a glass of wine she often jumps up onto my lap. But not for long and not entirely out of affection. From my lap, she can see more, and sometimes the tail suddenly flicks up to DEFCON 1, especially if a dangerous pheasant is sighted.

I often wonder what goes through her head. Does she live in an eternal present? She certainly has memory. When we walk down the Long Track to the bottom, Maidie sniffs the grass verges with great concentration. But when we turn to walk back up and round to the farmhouse, the little dog stops every twenty yards or so and turns, looking back to where we have been. Standing very still, she stares back down the track. What is she waiting for? What is about to appear? I suppose we are not so dissimilar, my dog and I.

23 May

On 23 May 1940, history was being made, lost and found. After German panzer divisions had smashed through French lines and raced into the valley of the Somme, they began to encircle the British Expeditionary Force and some retreating French troops clustering around Boulogne, Calais and Dunkirk. Defeat was inevitable, surrender likely. France would fall in only a few days and the evacuation of the BEF from the beaches of Dunkirk would begin. Shock and panic ricocheted around the walls of the War Office in London. Thirteen days before, Winston Churchill had replaced Neville Chamberlain as Prime Minister, and his famous call of 'Action This Day!' rang around Whitehall, and directives of all sorts flew out of Downing Street.

On the same day as French resistance was crumbling, its army surrendering en masse, the manager of the Commercial Bank of Scotland, 10 High Street, Selkirk, sat down at his desk. Each morning began with the same ritual, as he opened the post with his chief clerk in attendance. There was an unusual, official-looking communication, perhaps something that caused the manager to raise an eyebrow. A directive from the War Office required immediate action – this day! No quantities of paper or other combustible materials should be stored in the attics or upper storeys of buildings because they presented a greater risk of incendiary damage from the bombing raids that would surely come.

The stately offices of the Commercial Bank had once been those of the town clerk of the Royal and Ancient Burgh of Selkirk and the manager knew that there were many boxes of old documents in the attic and upper floors of the bank. An odd-job man was found and told to carry the boxes and sacks down into the back garden and burn the lot. Forthwith.

Next door to the bank was Masons' Tearooms, where Bruce and Walter Mason were bakers, making scones, teabreads, Selkirk bannocks, cakes and biscuits. Out of the bakehouse window, they saw plumes of smoke billowing upwards and went outside to investigate. Both were aghast to see four hundred years and more of the town's history going up in flames. Bruce and Walter pulled out some papers not yet alight and saw that they were from the sixteenth century and had royal seals attached. Rushing into the Commercial Bank, they pleaded with the manager for permission to go through the papers that were left and preserve what seemed to be important documents. This was summarily refused on the surprising grounds of confidentiality, a characteristic reflex of a bank manager.

Undeterred, Bruce and Walter resorted to covert action. Having given the odd-job man half a crown, they asked him to bring down the sacks and boxes of papers very slowly. Having quickly sorted through what had survived, they managed to save thousands of documents from the flames and secretly stored them in the bakehouse loft in flour bags, tea chests and barrels.

Perhaps they feared prosecution, perhaps they had breached some law or other. The Official Secrets Act? Even after the end of the war, the Mason brothers told no one that the documentary history of Selkirk and the farms around it was hidden amongst the bags of dried fruit, candied peel and wheaten flour stored above the tearooms. By the late 1950s, they decided that the coast was clearing and the brothers agreed to let Walter Elliot into the secret. The three men began the vast task of deciphering what had been saved. The documents turned out to be a revelation, as though a brilliant sun had suddenly risen over a dark landscape.

On 31 August 1531, the geography and pattern of ownership of our farm and the land around it was laid out in a document written out by the priest 'William Brydin vicar in Selkirk'. Philip Scot of Edschaw [Headshaw, a farm near Ashkirk] made a will in favour of his son, Robert, leaving him the farm that lies immediately to the south of ours:

His place commonly called Hartvodburn [Hartwoodburn] after his decease, the which place Philip Scot holds in tack of [rented from]our lady Margaret Queen of Scots lying in Ettrick Forest in the sheriffdom of Selkirk between the kirk lands of the most reverend archbishop of the diocese Glasgow, viz. Synton and Edschaw on the south and west, le Myddilsteid [now Middlestead] on the west, Haning [Haining] on the north and Selkirk Common on the east.

Astonishingly, all of the farms and properties still exist in precisely the same configuration. The major change is the subsequent passing on of ownership of the royal hunting ground (of which our farm was once part) of the Ettrick Forest and of the land owned by the Archbishop of Glasgow. Only thirty years before the Reformation swept over Scotland and broke up the great church estates, this document shines a brilliant light on the late medieval landscape around us.

Another document dated 24 January 1536 (or 1537, to take account of changes in reckoning the calendar) animates the map, giving details of rural crime, of accusations of unwitting reset. It reads like the legal residue of harsh words, perhaps even altercations. The accuser was our predecessor at Hartwoodburn:

James Bradfutt, Baillie of Selkirk showed how it was alleged by William Scott in Hartvod [Hartwoodburn] that the same [William or John?] was selling stolen skins openly in the market having no suspicion the skins were stolen . . . and that the said William Edmont frequented the market of Selkirk . . . to sell openly his merchandise without suspicion . . . and it should not turn to him in prejudice and slander.

These extracts are from the earliest documents, written by priests and notaries from 1511 onwards. After the Reformation of the 1560s the same men became lawyers, in practice if not by qualification, because they could write and knew the legal formulae. They also dabbled in banking and insurance, as well as sorting complicated cases involving slander and stolen goods.

Before Walter Mason died in 1988, he asked Walter Elliot to take the documents for safekeeping. Having no immediate family, he feared they might find their way onto another bonfire. Some documents were so dry and brittle they crumbled to the touch, others were damp and mouldy. But all were precious. With thousands of records and handwritten books, Walter Elliot sought professional advice. Dr John Imrie, the recently retired Keeper of Scottish Records, spent two days meticulously going through the collections before announcing that they were of immense national importance, especially the sixteen Protocol Books. These were the core legal notebooks kept by priests and notaries from 1511 to 1668 and they recorded land transfers, disputes, marriage contracts, wills and inheritances, letters and commands from the king and much of consuming interest. Nowhere else in Scotland had continuous records of the life of a community survived in such detail and quantity.

What became known as the Walter Mason Papers were also important culturally. Written at first in late medieval Latin and early forms of Scots in shorthand (to save paper), they showed how the language changed and developed over almost three centuries, how land use altered and how communities, especially town, country and the aristocracy, interacted. But most striking of all, they give occasional voice to ordinary people, the vast majority who lived in and around Selkirk, those who are very rarely heard before the late nineteenth century. All of this, the very best, most pungent sort of history, was painstakingly transcribed from pages and notebooks, frayed at the edges, soggy with damp and mould, nibbled by mice and partly burned by English invaders, Border Reivers and the manager of the Commercial Bank of Scotland in Selkirk.

24 May

On the wall of my office hangs a framed letter from a king, signed and sealed by him, promising to protect our farm. In late 1542, James V of Scotland wrote to the Abbey of Kelso with a warning. Written in Scots, here is a partial translation:

> Baillie of the Regality of Kelso and your deputies. It is our will and we charge that ye, bie yourself, your kin, friends, servants and portioners, not nae others that ye may let make onie invasion, skaith [damage], harm or displeasure to our burgh of Selkirk or inhabitants there of bounds and free-doms of the same in onie ways in time coming bie eating and destruction of corn sown or other ways [otherwise] but bie order and process of law discharging you thereof notwithstanding onie other of our private writings in the contrary. Because we have ordanit and command . . . to ablat(?) on you in our . . . in this time of trouble as you be required . . . Given at Edinburgh the last day of August and of our reign XXIX years.

The king's reference to troubled times reflected more than trouble. Earlier in August, Henry VIII of England had sent a small army north to prick James into supporting him in his break with the Pope in Rome. But the Scots king remained resolutely Catholic, burning Protestant martyrs below the walls of Edinburgh Castle. A few days before the letter was sent to Kelso Abbey on 31 August 1542, a battle erupted nearby. At Hadden Rig the Earl of Huntly scattered a small English army, but this success was blighted by farcical failure further to the west. At Solway Moss a huge Scottish force was defeated by a much smaller and much better led English army. Only a month later James V turned his face to the wall and died. He was only thirty years old.

From reading through the thousands of entries in the Protocol Books and piecing together their patchy geography, it is clear that the Deer Park and all of our grass parks in the East Meadow and near the farmhouse, down to the boundary of the Common Burn, formed part of Selkirk's South Common. And since the Abbey of Kelso owned the land immediately to the east, it seems likely that it was our corn that their men despoiled. The Haining and the southern part of Hartwoodburn Farm, what lies on the far side of the old Roman road, appear to fall outside the common. So it was not actually our corn that the Kelso men stole, it belonged to the burgh of Selkirk. Which is still annoying.

25 May

After another warm and restless night during which I heard heavy rain, the morning scents were intense. Not only were there the sweet pleasures of the stand of poplars at the corner of the Haining Wood, there was the lush and earthy odour of burgeoning grass, especially on the margins of the tracks where it is not grazed. And in the thorn hedge a single dog rose had opened. Its perfume was pungent, magical, a thing worth getting out of bed for.

Walking out to the East Meadow to check on the mares, the mini Shetlands and the Old Boys, not something that needs to be done so much in the summer, I could see they all had plenty of grass.

When Maidie and I came back up the track by the meadow, our neighbour was shaking out bags of hard feed for his sheep and cows. That surprised me. Perhaps he has stock that won't keep. The bull was with his cows and he greeted us with a mighty roar, more like deep trumpeting, that rang around the valley. And, my goodness, what a pair of balls that king of cattle has swinging under him.

26 May

Grey sheets of welcome rain blew down from the western hills this morning.

27 May

I feel we stand at a crossroads in our history. After years of ineptitude and muddle, our people long for strong or at least clear leadership to navigate us safely through hard times. Of course, the most pressing issue of all is the rescue of our dying planet. Everything else is a long way second.

My craven instinct is to turn inwards and look backwards, and it is difficult to resist. On the farm we do all we can in the fight to save our planet from ruin. We have planted three hundred trees, use no pesticides or artificial fertilisers, avoid long journeys, have put up thirty-three solar panels and with our mile of hedging encouraged the return of many animals, principally birds. They feed on the flies that in turn feed off the horse muck. We could no doubt do more, and it feels like a piddling effort in the face of China belching out kilotons of foul pollution every day, but what other option do we have?

On this grey, cloudy and cheerless morning, I confess my spirits are low. I fear for the future of my children and grand-children, and have perhaps only ten years of relative health when I can attempt to achieve something that might help them. But I am by nature an optimist, and when my mood lifts we should think about how to plan for an uncertain time ahead.

28 May

When walking is metronomic, the steady planting of one foot in front of another, thinking but not thinking, it can feel like therapy. This morning I try to walk off my low spirits and, after a couple of yanks of the lead, dragging her away from particularly pungent sniffs, Maidie falls in step. We follow the old track into the Deer Park and I begin to remember the past, stories my mother and my aunts told me.

She and her six sisters walked everywhere. Raised in Hawick and all of them employed in the textile mills when they left school at fourteen, they sought diversion from the thrumming, humdrum rattle and clack of the great looms, the dust and the deafening racket of machinery churning out bolts of tweed, cotton vests, underpants, and lambswool and cashmere jumpers. On summer evenings after supper they walked up the Teviot to its confluence with the Borthwick Water, paddling and guddling along its banks before gloaming gathered. At weekends, make-up on, hair done, cigarettes lit, they walked to dances at village halls, and walked home again through the light summer nights, teasing each other with stories of ghosties and ghouls and jumping when an owl hooted. And perhaps teasing each other about the lads they had danced with.

Every generation before ours walked. Time pressed less. Walking was cheap and, weather aside, under personal control. No one needed to wait for a lift or a bus. They just walked when they wanted to go and arrived when they arrived. With much less traffic on the roads, and often in cheerful company with a great deal to discuss after the dances, the miles flew by. Now we drive, mostly on our own, or are driven. We sit rather than stand. That seems to me to be a sharp shift in cultural habits, perhaps one that is less understood than it should be. Before we drove down our ancient tracks, three or four hundred generations walked them. And they saw, felt, smelled, intuited and experienced the land in a way that we may be losing.

29 May

Streamers of sunlight filter through the morning mist, light the landscape yellow and are then obscured. Grey, yellow, and then grey, like a stagehand rotating a coloured wheel of gels against a lamp. What will the day bring?

Eventually the sun wins the unequal contest and Maidie and I wander out for another brisk old-fashioned walk, instead of the usual halting wander.

I have been thinking a good deal about the dead. John Goodall, my Latin and Greek teacher, died ten days ago and this afternoon I will speak at his funeral. Wanting very much to do him justice, I will learn my lines and I hope I do it properly. He would expect nothing else. 'Speak up, boy!'

30 May

The wild raspberry canes by the Bottom Track have begun to leaf out and soon the little white flowers will come. It is a warm, damp morning, perfect growing weather, and now that the May trees' blossom is all out, I can cast my clout of a jacket.

At Windy Gates I stop, my attention caught by the eastern horizon, the high ridge of the Deer Park. Between the familiar trees, I see something that is not right, was not there yesterday. Near the stand of majestic beeches, all their dark foliage burgeoning, there is a strange silhouette. Not one of the saplings we planted, not a post set up by the gas pipeline people who are always pottering about on our land without asking. I can make nothing of it.

And then it moves, turning slowly towards the town, and the man raises an arm. He stands at the highest point of the ridge, where there are the longest vistas in all directions, and waves slowly from side to side, like a signal. Standing still, staring, I feel myself slipping through a crack in time.

In 1513 conflict crackled along the frontier between Scotland and England. The terms of the Auld Alliance were always vague, but they persuaded James IV to attack Henry VIII when he attacked Louis XII of France. A sequence of events was set in train that would end in catastrophe. In the late summer of 1513 the Border towns were set on a war footing. Rescued from the flames outside the Commercial Bank during another war, the Burgh Court Book

> ordains that night watches are to be kept by men and not boys; they were to walk on the backlands [behind the town's houses] within their watch and not go to potation and drink from nine o'clock to cockcrow.
>
> A watch of eighteen men, neighbours and householders, fully armed as best they might, had to walk each night . . . Failure to do so meant a fine of 12 pence.

The southern boundary of the town, part of the route of the Night Watch, is our northern boundary, the edge of the Deer Park. But it is low ground, next to a place still known as the Bog or the Bogheid. The Night Watch will have sent a man to the summit of the Deer Park to stand lookout, perhaps taking turns as they walked their circuit. It is inconceivable that they would not do this.

The Burgh Court Book for Selkirk, also rescued by the Mason brothers, vividly describes rising international tension, as James IV prepared for war with England and began to raise a vast army, the largest ever to march out of Scotland.

> 2 August, 1513, Finds and ordains all neighbours and indwellers to be armed for war after the tenor of the king's letters produced at the last wapinschawing [muster] to give their demonstration and show thereof in the Bog before the baillies [town officials] on Wednesday, St Laurence Day [10 August]. And that all indwellers for the weal of the town and country, having servant men and children, that they be produced at wapinschawing in the best way they can with one spear, lance and bow . . . To be fulfilled under the unlaw [fine] of 8 shillings.

Also finds that the neighbours about the hill [perhaps the Deer Park] lend their horse to bring 5 sledful of turf and who has no horse to come himself and give his pains for casting and laying of turf; and that each indweller send on servant betwixt this and Sunday to the lochend and places about the Bog where need is.

They built the turf wall precisely on our boundary. A tumble-down drystane dyke tops it now, but it sits on a narrow, raised ridge and the remains of a shallow ditch can be seen on our side. It is a monument to panic. The wall reaches as far as the edge of the Haining Loch, but in reality it was a flimsy gesture and would not have delayed a determined attack for very long. Mercifully, it was never needed.

On 9 September 1513, James IV lost a battle he should have won and died in the ruck of hand-to-hand fighting. Flodden was a disaster, weakening the Scottish state profoundly and ushering in a century of violence and disorder on both sides of the border.

Perhaps the dog walker I saw greet his friend on the Deer Park ridge knew that he was moving through a rich landscape full of centuries, full of hidden incident. But what is certain is that he sought the highest point so that even on a cloudy morning he could look out over the land. Pleasure and not anxious necessity had taken him to the high vantage point.

Perhaps I think about the past and the dead too often, but yesterday I managed to say some resonant things about my old teacher, John Goodall. He gave me and others many gifts and I was glad to thank him.

31 May

The secret scents of the earth swirled in the morning breeze. After two days of warm rain, growth is everywhere. On either side of the Long Track, the grass parks are now knee-high; on its verges, tall stalks of cow parsley seem to have sprouted overnight, their lacy flowers casting a bitter perfume. Yesterday my neighbour ploughed much of the Tile Field, obliterating the spreading contagion of marsh grass, as the chocolate furrows folded over. The deep loamy smell drifts over the grass parks as the world unfurls its fertility. When the sun comes, the land will glisten.

June

1 June

Early on a bright morning, the cows were lying in wait for the sun. In that strangely swinging gait that suckler cows have, they had plodded up the Long Track from the Doocot Field and lain down on the east-facing slope of Huppanova, where the warmth would first be felt. As Maidie and I clanked through the metal gate, none moved except to turn their great heads to see that we were no threat. Most of the calves lay beside their mothers and few bothered to look up as we passed.

Out in the East Meadow, I could see three of the Old Boys grazing but could make out no sign of the oldest, Gem. In the middle of their seven-acre field are three stands of tall nettles (which I must knock over soon) but, as we walked down the track to the Deer Park and closer, I thought I could make out a fourth. Black nettles? No. Gem. Like the cows, he was lying with his legs tucked under him, but unlike them his head was also down. Now thirty-one, he is a Very Old Boy, and, having tied Maidie to the fence, I walked over to check on him. His eyes were open but he seemed immobile, stuck. Despite rocking him from side to side at the withers, he would not or could not get up. This is what happened to lovely, elegant, arthritic Murphy in the winter before last. His legs failed him and he never got up again.

Thinking I should have brought my mobile (but are there no places or times now when we might be free of the need or means to communicate?), I rehearsed what I would say to Lindsay. When I turned to walk back to pass the other three horses, they began to follow me, thinking I might have brought bowls of hard feed. And then, thinking the same, Gem got up. Very quickly and easily. Old fraud. Lindsay told me that when he sleeps Gem now snores very loudly. My wife will probably read this, and so I will make no further comment.

2 June

Maps fascinate me because they record history as well as geography. I have four faded old maps of our farm and the land around it, and I sometimes pass a happy hour poring over them, comparing them with each other and the land as it looks now. The youngest is the Ordnance Survey of 1900 and while the oldest is undated it looks as though it was drawn in the late eighteenth century. The detail is humbling; it must have been the work of many hours to plot and mark even individual trees, as well as small ponds, old quarries that are now nothing more than scoops, and tracks that are little more than sheepwalks.

Reflecting the fact that land was wealth and the sole source of food, the Selkirk Protocol Books brim with documents confirming, disputing and denying its ownership. And when the old maps are read with the documents of the sixteenth century at my elbow, patterns emerge, stories unfold.

In the desperate decades after the disaster at Flodden in 1513, Border communities struggled to survive, scratching a meagre living from farms frequently plundered and despoiled by armies, forces of skirmishers and the bands of horse-riding thieves who later acquired a dubious, dark glamour as Border Reivers. In 1536, an entry in the Protocol Books outlined the boundaries of Selkirk's South Common. It was vast, perhaps five thousand acres, and it included the Deer Park as well as all of the other grass parks and paddocks on our farm.

But its bounds were difficult to maintain and police. Surrounding lairds were constantly encroaching, shifting dykes, attempting to assert ownership by cultivating large parts of the ancient common and initiating expensive court cases to back their spurious claims. It was during one of these disputes that Provost John Muthag and Baillie James Keyne were ambushed and murdered on their way to plead at the Court of Session in Edinburgh.

At its zenith, the Selkirk Common had once extended over a huge area, approximately eleven thousand acres to the north and south of the town. By 1517, only two years after Flodden, the burgesses had decided to ride around the marches of their land, the beginning of an immense tradition. But by the middle of the sixteenth century, the town's ownership had slackened dramatically. Rather than struggling to hang on to such a vast acreage, much of which the townspeople could never work or use, the burgesses were granted a thousand acres by James V. They could cultivate it or lease parts of it. The small common was divided into three farms: Linglie in the north, and South Common and Smedheugh in the south. On my oldest map these boundaries are clear, and they are still visible in the land-scape.

Sadly, our farm had slipped out of Selkirk's ownership and the boundary of our long East Meadow is marked by a crumbling, lichen-covered dyke (so tumbledown that I have had to have a fence erected in front of it) and on the eastern side the fields of South Common begin. Our land had become the property of the Scotts of the Haining and our story becomes bound up with theirs for almost four centuries.

But memories are long, and next week Selkirk will ride around the marches of its old common to the north of the town to make sure that no dykes have been moved, no lairds have encroached and Selkirk's ancient rights to its land remain.

3 June

Silent witnesses in the landscape, dormant, budding, shedding leaves through hundreds of years, hardwood trees live much longer than men or women, often outlasting buildings, sometimes dominating a skyline. When the burgesses of Selkirk and their supporters rode the marches of their common, they sometimes deliberately marked old trees that stood on the margins of the town's land, scoring their bark, using them as boundary markers.

On my old maps, individual trees are often plotted with great precision. The earliest map, probably dating from the 1790s, shows only the southern slopes of the Deer Park, but it marks two hardwood trees and surrounds them with what might be gorse. Two hundred and thirty years later, they are still there, clinging fast to the limestone, their roots finding cracks and fissures below the thin soil. They are ash trees and long-lived for their species, but very close to the end of their span. In last summer's devastating storms, the westernmost ash broke in half, losing much of its upper trunk and top canopy. When I looked at the heartwood, it was black, rotten, having been attacked by a fungus of some kind. This summer I will log the wreckage but leave what is still upstanding. The lower trunk and the main limbs are still alive and now in leaf. It presents no danger, so I won't fell it. The old tree can take its own time to die.

All of the old maps show what must have been a glorious avenue of hardwoods on either side of the Long Track. Again there is precision, with the mapmaker plotting twenty-two on the eastern side and twenty-four on the west. The Ordnance Survey of 1961 to 1981 marks some as still standing, but the farmer who sold us our land cut them all down about thirty years ago before he left, grubbing up the stumps to maximise his acreage, destroying something much more valuable and beautiful. There is one dogged survivor, a sycamore that grows in the verge of the track. It has been hacked at several times but always comes back. I call it the Life Force. Sycamores can live for two hundred and fifty years and the two survivors in the Top Wood are certainly ancient, perhaps part of the original planting. They are much scarred by storms but majestic, defiant.

The Ordnance Survey of 1900 shows all of our farm and the Deer Park as policies, a managed, even manicured landscape around the Georgian mansion house at the Haining. The sole purpose of policies was to give pleasure, create eye-filling vistas and also keep the rude mechanicals who worked the land for a living at a discreet distance. Horses or even decorative, shaggy Highland cattle grazed the policies to keep the grass manageable. What are now productive parks were dotted with many individual trees and several stands, and patterned with tracks that were used as carriage drives, routes for an afternoon excursion for the ladies and gentlemen of the big house in fine weather. It is all very different now. The rude mechanicals have taken over.

4 June

On the south-facing wall of the second floor of the farmhouse are three large windows that look out over a panorama of the Tile Field and enough of the sky to allow a quick judgement on the weather as I walk past. But this morning I stopped for a longer look. In order to stem the tide of marsh grass, our neighbour first dug several deep pits where the ponding of rainwater, even in dry spells, had given away trouble underground, the location of broken drains. He told me that in the thick clay some had sunk four feet down and were hard to find. Yesterday he ploughed up the parts of the field covered in marsh grass, and lit by a slanting, dawning sun, the chocolate furrows and the lime green of the unaffected acres was a stunning contrast, so sensual I felt I could taste it. Like the purple young leaves and the green buds of the acer by the stable yard, here were two colours found together in nature that would adorn the palette of a weaver or a fashion designer.

Today he is harrowing the furrows and making a seed bed for new grass. To suppress the regrowth of the marsh spikes, he will want it to establish quickly and, with the rain forecast for the week, we should see a first, faint flush soon. One of the reasons we have so little of this horrible, useless stuff is that all of our pasture is now well established. None of our paddocks have had a plough through them for twenty years and in the Deer Park probably not since the Napoleonic period of the early nineteenth century. With a liberal mixture of some weeds, herbs and wild-flowers, the old meadow pasture has what farmers call 'a mattress' under its top growth, a good hold on the soil gained after many years of non-disturbance. And it is in my view the best sort of pasture. The mixed content gives our horses very high quality fodder and a strong dose of health-preserving herbs and vitamins. Perhaps that is one reason why Gem and the Old Boys are living so long.

Around the time of the disputes over the Selkirk Common, farming was changing. The need to rotate crops was increasingly recognised, as was the need to shift animals around to get the fertilising benefit of their muck. This later led to a demand for quick-growing sown grasses that could also be taken as a hay crop. Not all grass is the same, or, on closer examination, looks the same. Most of the new, quick-growing seeds were a mixture of cocksfoot, a pasture grass that develops purplish spikelets when left to grow tall enough to be cut for hay. Timothy is named after Timothy Hanson, an American farmer who developed it in South Carolina in the 1720s. It is also known as Meadow Cat's Tail and its flowerheads are very easily recognised, pale grey, shading to beige. Ryegrass is thicker bladed and found on lawns and cricket pitches, while clover is technically classed as legumin-ous or vegetable-like. Sweet vernal grass is often mixed with ryegrass and it is the variety that gives off the glorious smell of new-mown hay. Timothy seems to grow best with us in our clayish soil, but up on the Deer Park I have found clumps of all of the new grasses of the early eighteenth century clinging on to the limestone-enriched soil. We have not let the park for some years, and it desperately needs to be grazed to keep it manageable. But first we need about a mile of new fencing, and that will break the bank.

5 June

After a night of prolonged rain, sometimes so heavy that I could hear it drumming on the slates, we woke to a thick drizzle. June is when it should be possible to unwrap, but here we were wrapping up against the miserable weather, me in a hat and waterproofs and Maidie in her pink coat, cuffing at it with a front paw. Lush-leaved and dripping, the land looked like a temperate jungle after the monsoon, and in the warming woods of the Hare plantation on the far side of the valley streamers of mist were drifting up into the low clouds. Washed off by the heavy rain, the tiny white petals of thorn tree blossom fringed the puddles of the Top Track like lace collars.

History has tapped us on the shoulder once more. Where the Long Track turns towards the mound of Selkirk Castle and the town beyond, Rory has found more coins. Two carry the head of Edward I, one minted in London, the other in York, and both represent more buttressing evidence of a large army camp on 25 July 1301. A further find is a rare Irish coin with the head of King John, an instant cue for thoughts of Robin Hood, the Merrie Men and the dastardly Sheriff of Nottingham. What makes these muddy coins sing of the past is not only the context we have been able to assemble around them, but something more magical. Before Rory dug them out of the soil, the last pair of hands to touch them belonged to the man who lost them seven centuries ago.

The fields have given up more secrets and the pieces of a jigsaw are beginning to make a picture. Having found an ancient weight from a set of balance scales by the side of the Long Track, where it crosses a stream, Rory has found another relic of medieval business, of bargains struck and hands shaken. His metal detector buzzed loudly when he uncovered a lid from an early set of cup-weights. These were used to weigh bullion and other valuable items needing precision. On the lid is the French fleur-de-lis design. Did this belong to a merchant who had travelled a long way to the borders of Scotland? Ghosts are walking through this morning's mist.

6 June

Lit by a hazy, hesitant sun, the subtle colours of the temperate jungle are glowing this morning. Perhaps most striking are the perfectly ridged leaves of the many beeches in the hedge by the Top Track, and their soft and quiet beauty is counterpointed by the yellow lichen on the branches. As Maidie sniffs the long grass on the verge for last night's mice, voles and rabbits, I take my time and remember summers long ago when I was surrounded by loud, neon, wholly artificial, glorious technicolor.

In the school holidays, as young as fourteen or fifteen, long before there was any legislation to prevent it, I worked as a second man. In the middle of Kelso there was a wonderful Willie Wonka factory that made lemonade. In the main building a clinking, clunky production line of empty glass bottles shoogled unsteadily around a network that looked like a giant version of a model railway until they arrived at the spoots, the point where they were filled with fizzy, vividly coloured lemonade of apparently endless variety. Labels were then gummed on for Limeade, Raspberryade, Cherryade, Orangeade and many other ades. There was American Cream Soda, something called Palletta (very like Limeade) and, the biggest seller of all, colourless Plain Lemonade.

In summer the Borders worked up a great thirst for these sweet concoctions and, as a second man, my job was to help unload the heavy wooden cases of lemonade from the delivery lorry parked outside cafes, shops, pubs and hotels. Our products were very popular – and very good – and we seemed to make weekly journeys with repeat orders, picking up crates of empties in an age before toxic, throwaway plastic, and replacing them with the neon colours from the factory. Splits were popular at agricultural shows, weddings and summer functions. These were small bottles with metal caps that could be levered off and the nectar sucked out with straws.

The secrets of the lemonade factory were kept in the still room. A loft space reached by a long, wooden staircase, it was presided over by two ladies in white laboratory coats. From large, squishy containers of concentrated flavours which had probably never seen a raspberry, an orange or a lime, they mixed potions in the right proportions. These magical mysteries were master-minded by Margaret Allen, a great beauty with classic 1950s film star looks and immense, genuine charm. Always with impeccable make-up and lipstick and maybe a hint of perfume (or was it the concentrate?), she dazzled the awkward, gawky teenagers on the factory floor, me included.

Just like the smell of newly cut grass, white marquees and men in shirtsleeves, for me lemonade means summer – preferably warm lemonade, drunk through a straw.

7 June

Driving to Berwick-upon-Tweed for an early London train (that turned out to be late, as usual), I passed through the tiny hamlet of Carham, a few hundred yards on the other side of the English border. No more than eight houses, a church and a farm steading, it still has a red phone box by the roadside. As I slowed down I noticed that irony is alive and well in North Northumberland. The red phone box had a sign on it: *Carham Visitor Centre*.

8 June

Last night and this morning's heavy rain followed a fortnight of mostly wet weather. At the corner of the Bottom and the Top Tracks, where the larch, the Norways and the American red oaks were parked, there is a riot of fecundity. As each tree reaches upwards to compete for the light, there is doubtless an invisible underground wrestling match going on amongst the intertwined roots. In the damp darkness all are snaking backwards into the rich, loamy bank at the foot of the Top Wood ridge. There they will spread and suck in all the goodness they can in order to help with the struggle above ground. The larch is tallest and this morning I could see hundreds of new lime-green cones bursting with vigour on its elegant branches. They are perfect, these triumphs of natural symmetry, each one apparently identical. I like the chaos in the corner, and all of the competitors seem to be thriving more or less in this perfect growing weather. But now we need some sun and warmth.

9 June

And this morning we have it: sun, warmth, the ground drying in a gentle breeze and the grass growing before our very eyes. Such scenes of peaceful fertility were often little more than a fond wish in the Borders five hundred years ago. I have been looking through the Protocol Books preserved by the Mason brothers and they make grim reading; inevitably there are lists of disputes, and of ruin and death in the century of warfare, and raiding that disfigured the landscape after the disaster at Flodden in 1513. But occasionally there is a smile, a flush of recognition, a document that springs off the page, one that speaks pungently of the texture of the old life in the Border countryside.

I would like to have met Gibbie Hately. He was a minor land-owner and farmer who lived in a peel tower at Gattonside, near

Melrose, about eight miles from here. A protection against the raiders and thieves who disfigured society for almost a century, many peel towers were built in the sixteenth century, and behind the barmekin wall around them people and stock took refuge when raiding parties struck.

Made in 1547, Gibbie's will is distinguished by a clear sense of a life lived with relish in the first half of the sixteenth century. After reading it I felt I would recognise him leaning over a five-bar gate, looking at his lambs of a summer evening, and I would have liked him. The original document is written in wonderfully expressive Border Scots, the language Gibbie Hately spoke, and it is speckled with words and expressions long lost, so here is a translation:

To Geordie Basten, for the great trouble he took with my plant land when I could not attend to it myself and the expensive drive to the market of Stirling for which he could not be prevailed upon to take anything – no, not so much as the price of a single thousand of plants [probably kale]; to him I leave two mounds of turfs [peats], two rows of drying peats from Rob's bog and a lypit-spade and a flaughter-spade [both peat-cutting implements] for cutting the same.

To Patie Dickieson for his kindness and attention even though he had gotten a thumb cut off at Elwan Bridge [near Lanark] by his brother in a duel; despite this, he had his men sow the Cotland barley and the broom seed on the face of the brae, the plants in the Abbot's Meadow and a few oats in the east corner of the Quarterland and a capful of linseed [to grow flax] in the Harper's yard; To him, I leave an oat riddle with the iron rim, my three best weights and the broom seed basket, my fish spear and my fishing tackle.

To Andrew Fisher of the West Houses, for helping me

when I fell into Hamilton's Burns with holding the Quaich too often to my head [getting drunk] on the Stears [?], on the Thursday evening of a fair day. I leave him my hazel staff with the horn head, my best bonnet and hazen [stockings] and the new shoes that Willie Fair brought me from Sandy Inglis of Selkirk, made of good buck's hide and the soles of the same made from [the pelt of] the big boar shot by the Laird of Faldshope; also my farming oozlles [utensils], and snuff-horn, trimmed with silver.

To kind Adam Ormiston, the hangman of Edinburgh, for helping my father out of prison the night before he was to be hanged for killing one of the king's deer on the Cauldshiels Moor and the king's forester of the Melrose end of the loch who was very keen to make him his prisoner for killing the beast he had no right to. To him, I leave my great-grandfather's silver tankard and one quaich which my great-grandmother received from Laird Maitland for helping nourish his brother Robert. Also my father's gold ring in which [is set] the emerald he promised to Adam Ormiston if he could slip him out of the window of the prison unseen, which he faithfully did for the love he bore my father.

To the Laird of Langhshaw, I bequeath my broadsword and my dirk. To the Laird of Hislop, all my hawks and hounds. To Laird Usher, my brother-in-law of Fastenfield, a hundred marks Scots and my riding horse and two older horses I took from the lads of the Border when they came one night to harry me. To my brother-in-law commonly called Longsword of Faldonside, I leave two hundred marks Scots. To the Abbot and monks of Melrose [Abbey], I leave four hundred marks Scots, to pray for my soul and the welfare of my son, Jock. To Jock I leave a thousand marks Scots, one Cotland [about five acres] and one Quarterland [26 acres, a quarter of a husbandland], the Abbot's Meadows and the old peel [tower] which I hope in God he will keep from all the English loons as his forbears have done before him.

This was clearly a will made during what are known as the Riding Times, a long period of raiding and warfare that lasted until 1603 and the Union of the Crowns and even beyond. The criminal society of the reivers coloured almost every codicil in Gibbie's last testament. But the final sentence made me laugh.

10 June

This morning sunlight flooded the valley. It was still, cloudless and hot at 7 a.m. Beyond the old Roman road, the fields cant to the north as the contour lines climb up to the southern ridge and I could see that my neighbour had cut the lush grass park beyond Hartwoodburn steading. The green rows reminded me of braided hair. Three tractors rumbled past the bottom of the Long Track, each towing the machinery needed to lift the rows of cut grass and bale them for winter silage. Gleaming in the sun, the contractors' huge John Deere tractors seemed new. It cheered me to see the land produce its fourth crop of the year, after the spring lambs, the calves and our winter store of freshly cut logs.

This afternoon I hauled all of our old and not-gleaming equipment out of the Wood Barn. With great difficulty, I managed to re-inflate the sagging tyres of the quad bike and the grass cutter we use to top the paddocks, but I could start neither of them. All winter they have stood idle and unused, happed up in old horse rugs, and I suspected that the petrol left in their tanks had grown stale. I splashed in some fresh fuel from the jerrycans. Still nothing, dead as a post. A mechanic needs to be summoned, but I should avoid that by keeping the quad bike going through the winter with logging jobs. There are plenty.

11 June

Last night I walked a ghost road. Spear-straight, the Long Track points due north through the Doocot Field and then abruptly disappears into a wood planted across its line. It then seems to turn sharply to the east towards the Georgian mansion of the Haining, but the trees and their planting told me this was a later diversion. Beyond the darkening wood, at the foot of a sloping field where an unexpected clump of purple rhodendra grew, I could see the line of another road. It seemed to me that the Long Track had once joined it. Like a holloway, it is steeply embanked on either side and it aims west to the Ettrick Valley. I could make out the evening shimmer of the river.

The western section is broad and a place where carts could pass each other. But on the road-bed there stands a line of mature hardwoods: oak, ash, sycamore and beech. By their girths and the toppled debris of even older trees that once stood alongside them, I guessed they were planted at least two centuries ago. The old estate map of the 1790s shows a line of hardwoods but no road.

As I walked through the long grass east towards Selkirk, I could see that the line of the holloway ran directly towards the West Port, the gateway into the town. Modern housing has obscured any junction. Walter Elliot told me that I had walked all that remains of the old medieval road that runs west up the Ettrick Valley and that it probably fell out of use in the 1770s. The trees agree with that judgement.

In 1757, John Pringle, a merchant who had made himself wealthy in Madeira, came to live at the Haining. He extended the policies, planting woods and creating gardens, and at that time he may have wished to move the old road into Selkirk further to the north, where it runs now. Once again the rude mechanicals were kept at a distance. And, once again, they were returning to rediscover their history.

With his metal detector and instinct for the lie – or the truth – of the land, Rory Low has confirmed the nature of the road. His discoveries have made it come to life. A Henry VIII sovereign penny (so-called because it shows the king enthroned) and an Elizabeth I sixpence are surely the first of many coins to come up out of the road-bed as he sweeps its line. On busy market days, farmers led carts and drove animals, the high banks keeping them safely corralled, others walked, carrying their produce on their backs, and some rode. And occasionally they dropped their hard-won coins. The Protocol Books add atmosphere. In Selkirk's taverns and ale-houses, farmers took a drink after the bargaining in the marketplace and the baillies, the town officials, regulated the price of ale by chalking it on the doors of each hostelry. And in an age long before standard measures, they agreed on the size of the Selkirk Stoup, a jug of ale. Maybe some, like Gibbie Hately, had to be fished out of the Ettrick on their unsteady evening journey back up the valley.

When I walked down this old road, beside its ghosts, their chatter blowing on the freshening breeze, I noticed something odd. Where the banks are highest, there was a bend around what seemed like a grassy mound. Rory is much intrigued by this and wants to get the long grass cut so that his detector can find a clearer signal. Was this a tower like Gibbie Hately's? It would have commanded long vistas west up the valley and over the hills beyond.

It was gloaming by the time I walked along the Top Track and I could see the lights in the farmhouse kitchen twinkling.

12 June

A wild north wind riffles the long grass like the waves of a choppy sea, and the new leaves are turned inside out, showing their light undersides. The skies are dark, rain threatens and it is cold.

13 June

Our book festival in Melrose began in a steady downpour that persisted until 7 p.m. There were wonderful sessions with Neil Oliver and Kate Humble, but neither were even close to full and that vexed me. In the rain, people don't want to turn out and sit in a damp marquee, no matter how warming the words of these brilliant people.

14 June

This morning the world shifted on its axis and slipped once more through a crack in time. Two miles to the north, on the flanks of Peat Law, I saw three hundred and more riders led by a standard bearer, his flag streaming behind him in the breeze. Lit by the streaky sun, the grey horses stood out as this cavalry force climbed the hill to the Three Brethren. Cairns that mark the marches of the North Common, they were built centuries ago on the summit of the hill. There, the riders stop and sing, remembering a time out of mind. 'Hail Smiling Morn' rings out over the glow of the heather hills. 'At whose bright presence, darkness flies away, FLIES AWAY!' For at least five hundred years, men from Selkirk have saddled their horses to patrol these uplands and defend their rights and their common. From Windy Gates, I could make out a long, streekit line zigzagging before turning north to the cairns. For a long time, I watched history come alive, much moved by the sight of the largest mounted cavalcade in Europe.

It is Common Riding Day, the first Friday after the second Monday in June. Last night, the traditions began to roll back the years when the Burleymen walked the streets of the town. Burley refers not to stature but to statute. It is the phrase 'Burgh Law' rubbed smooth by the centuries. The Crying of the Burley ends with a proclamation, a call for the townspeople to assemble in the morning to ride around the marches or support those who will. It ends with a stirring exhortation that carries the hint of a threat: 'There will be all these, and a great many more and all will be ready to start at the sound of the Second Drum.'

They begin early. At 4 a.m. the Rouse Parade of flutes and drums tours the town to wake the Standard Bearer and the Provost, and everyone else. Two hours later, after much ceremony, the First Drum sounds and all are out on the streets, dressed in their best, bedecked with rosettes, and marching. Linking arms in long lines abreast, the foot parade stops at certain places and sings songs that are only ever sung there on Common Riding morning. It is the soundtrack of centuries of continuity and a community coming close together to remember its shared history, celebrating nothing more, or less, than itself. At 6.30 a.m., on a narrow balcony, the burgh flag is 'bussed', blessed, the Standard Bearer installed with a red sash, and awkward, time-honoured Victorian phrases uttered. Reading from a script taped inside his bowler hat, the young man who has just been appointed to lead the cavalcade promises to ride the marches and return the flag to the Provost 'unsullied and untarnished'. It never is, even on a rain-soaked morning.

When the riders return, clattering up the old toll road into the town, something unique and magical takes place. Flags are everywhere, those of the craft guilds, the ex-servicemen, the exiles. On a dais in the market square more than a thousand people watch as all of the standard bearers cast their flags. To the accompaniment of the silver band, each man performs a similar ritual. Planting his feet apart for balance, he begins to wave the flag slowly from side to side, then behind each shoulder, pulling it forward. Squatting down on his haunches, he then rotates the flag in wide circles over his head before standing up to perform the last movement as music stops. It is spine-tingling, very moving, stirring ancient resonances, and it is seen nowhere else.

After all the flags are cast, the band plays a lament, the 'Liltin''. A friend once whispered a question during the two-minute silence that followed, 'Is this for the fallen of two world wars?' 'Yes,' I said, 'but also for those who fell at Flodden.' Five hundred years ago. Memories are long and do not fade. When we came to live on the farm permanently, I went to the casting of the colours with an old friend I had played rugby with. Craggy, hard-bitten and no-nonsense, he had tears in his eyes when the 'Liltin'' was played. 'Aye,' he said, perhaps to himself, 'we come from nothing small.'

15 June

Late back home from the joys of the book festival and a day when the sun had shone, I drove in the half-dark of the summer up the Long Track. Suddenly a big, very white barn owl lifted into the air. Flying first to my left over the grass park, it crossed in front of the headlights, only a few beats of a metre-long wing-spread taking the great bird higher. The owl then flew up to my right before circling over Windy Gates. It was playing with me, and welcoming me home.

16 June

Today is my sixty-ninth birthday, a number I find hard to credit, never mind absorb. Perhaps I shouldn't. It is only arithmetic and I am blessed to be alive and in reasonable health, despite my serial excesses. My dad died of a heart attack when he was barely seventy and if I pass his mark I shall be happy. On the way up the Bottom Track with Maidie, I noticed three purple foxgloves, their trumpets beginning to open in the morning light. I am lucky to be alive.

17 June

On a blustery, puddle-splashing morning, my three-year-old grand-daughter knocked on the window of her house as I passed. Pulling on her pink wellies, pink hat and flower-covered waterproof, she wanted to walk the Long Track with me and Maidie. About halfway down, I pointed to some animal tracks in the mud. Perhaps one set might have been roe deer, I explained, as we hunkered down like North American Indian trackers, and the other might have been badgers snuffling about in the darkness before dawn. 'No, Bada, polar bears and penguins.' If the climate keeps changing rapidly, the wee one might come to be right. Reindeer instead of roe deer.

Moments later, mostly to herself, Grace began to sing 'Twinkle, Twinkle Little Star, how I wonder what you are'. As she walked, doing all of the actions, singing the first verse over again, I don't think she saw her grandpa, her bada, pull his hankie out of his pocket.

18 June

In the Doocot Field, more mystery has risen up through the grass. Rory Low has found a strange sort of lead shot. Spheres larger than a musket ball have been cut into quarters like segments of a deadly orange and it does not seem as though they have ever been fired. Perhaps they were intended as fragments of canister shot, a kind of shrapnel fired from cannon in the seventeenth and eighteenth centuries. It was used to devastating effect at the Battle at Culloden in 1746, as the clansmen charged the government artillery. On a very misty morning, out with Maidie on the Long Track, it occurred to me that this mystery might have something to do with mist and the quartered lead shot was the ghost of a grisly story.

In the War of the Three Kingdoms – thoughtlessly, inaccurately and chauvinistically known for years as the English Civil War – Scots played crucial roles and one of the turning points of the long conflict took place as cavalrymen crossed our farm. Charismatic and tactically brilliant, James Graham, Marquis of

Montrose, had formed an alliance with the MacDonald clan leader Alasdair MacColla and together they defeated the armies of the Covenant, the allies of Oliver Cromwell and the Parliamentary Party in six battles, most of them in the mountains of the north. But by the late summer of 1645 the clansmen had decided to follow MacColla westward to attack Clan Campbell lands and Montrose was forced to march south to the Borders to raise more troops loyal to Charles I. The recruitment drive was not successful and, to make matters worse, a large Covenanter army had tracked the royalists' movements.

Early in the morning of 13 September, our little valley was blanketed in a dense mist. To the north of Howden Hill, on the flat ground by the Ettrick at Philiphaugh, Montrose's shrinking army had camped. Only about a thousand strong, they had dug ditches to defend their position and their officers had billeted themselves in Selkirk, about a mile away. All were entirely unaware that Sir David Leslie's army was approaching fast from the east, hidden by the mist. Royalist scouts reported no enemy activity, presumably because they could scarcely see more than a few yards in front of themselves. Mist muffles sound, but if any royalist riders had ridden down the Long Track from Philiphaugh, they would surely have heard movement along the old Roman road.

Suddenly and without warning, Leslie's forces bypassed Selkirk and charged in a frontal attack on the royalist position by the Ettrick. Montrose was roused and he rode hard from the town to marshal his meagre forces. At first his Irish regiment repulsed assaults, but neither they nor their commander realised that Leslie had sent two thousand cavalry to outflank them and attack from the rear.

On the old Roman road at the foot of the Long Track, unseen and unheard in the morning mist, their pistols loaded, their buglers silent, the squadrons of horsemen cantered over the still-hard surface. Sound carries over Howden Hill and they will have heard the thunder of battle on the other side, the shouts of men, the discharge of guns and the clash of steel. Hidden by the hill, they cantered down our little valley, turned north and formed up at the foot of the steep slope. Then they splashed across the Ettrick, kicked their horses into the gallop and their charge roared across the fields. Taken in the rear, the royalists were attacked from all sides, rolled up and routed. Surrounded by thirty of his own cavalry, Montrose cut his way out of the encircling Covenanters and fled into the hills.

Col. Manus O'Cahan's Irish regiment was persuaded to surrender in return for their lives, and then immediately betrayed. 'Jesus and no quarter!' was the baleful, bloodlusting cry and, at the insistence of the Covenanting ministers, the popish Irish were cut down, as were three hundred camp followers, many of them women and children. The nearby place name of Slain Men's Lea remembers this cruel, cynical and soulless act of senseless slaughter. Following a perverted logic, these clerics thought they flew on the wings of Heaven, doing the bidding of the Lord by ridding what they called the Godly Commonwealth of these idolators, but in any recognisable reality they were, in fact, barbarous butchers. As with Isis and the Taliban, the distance between what they believed they were doing and the horror of what actually went on was and is unbridgeable.

Perhaps Rory has found the remains of a skirmish in the mist. Royalist scouts did engage some of Leslie's outriders and perhaps the quartered lead spheres were dropped in a melee. Soldiers did carry spare lead to cast their own musket and pistol balls.

The morning of 13 September 1645 was not the last time history rumbled across our farm and its fields, but in the centuries that followed much less blood was spilt on its stones.

19 June

Very surprisingly the Protocol Books are silent, making no mention of the battle and the slaughter that followed at Philiphaugh. Instead they continue to record the detail of domestic life. Many of the entries are written in a rich and vivid Scots, what was clearly the language of all who lived in and around Selkirk, masters as well as servants. Not until much later was social class given away by accent. I have translated much of the Scots but tried to preserve the syntax so that it is just possible, across four centuries, to hear people talking.

When Bruce Mason died in 1963, his attic was found to be crammed with books and objects of all kinds and from many periods. Jostling for space with Neolithic flint arrowheads, Roman pottery, Chinese snuff mulls and French glass paperweights were many books, some of them of great antiquity. Buried in the magpie hoard were two hundred pages from a previously unknown Selkirk Protocol Book, whose entries dated from 1557 to 1575. There were also folders of random pages that took the story of the town and the farms around it into the seventeenth century.

In the summer of 1569, Selkirk was simmering in iniquity. On 16 July, the Court Book entry reads:

> The inquest aforesaid finds that the provost and baillies does not their duties concerning their office in suffering a multitude of whoremongers, whores, and their common oppressors to remain within the town in respect that they were delated [reported – in the sense of being informed against] and ordains the said provost to put them out of the town according to their duty and if they suffer them to remain unpunished the said provost and baillies are in default thereof.

It is striking that a small community should have sustained not one or two but 'a multitude of whores, and their common oppressors', or pimps. Perhaps fewer than a thousand lived in sixteenth-century Selkirk and my suspicion is that on market days in particular more were offering their services openly than was thought seemly. Despite the glares of the local ministers and church elders.

In the decades following the Reformation of the 1560s, what became known as the Parish State had come into being, with the burgh and the Kirk overseeing almost every aspect of life, public and domestic. In January 1572, James Kerr was charged with 'lying in fornication with Janette Chisholm and the said James was bound and obliged never to have melling [intimacy] with her except he make completely and solemnly the holy bond of matrimony with her'.

Marriages were not only insisted upon, they were also patched up. David Stoddard was accused of evicting his wife, Margaret Scott, but he 'declared that he had never deported or put her forth from his house and likewise was ready to receive her and use her as his wife to his power in all agreement'.

Other vices were recognised but regulated. Gambling for money was discouraged:

> The which day [the entry is, in fact, undated] the whole community has ordained that no young men or other indwellers such as honest men's bairns or servants who have the credit of other men's goods play at cards or dice, except for ale, in time coming whether within the burgh or without under the pain of remaining in the tollbooth [prison] in irons or in the stocks.

Dated 26 May 1591, a strange contract was entered into:

> Thomas Kerr, writer [lawyer] in Selkirk promises his brother James in Whitmuir and John in Whitmuirhall that, from the feast of Whitsunday, 1592, he will not drink in any place in Selkirk or other places where he has to pay silver or money except in his own house where he is allowed three chopins [three Scots half pints, more than two litres] per day. Also

any drink in his work service with his master where he shall get his food and drink for nothing. If Thomas bides by the contract he will get the grey russet breeks [trousers] which James was presently wearing and the white fustian doublet which John was presently wearing.

Behind this entry there seems to lie a long story of broken promises, alcoholism and penury, as well as a society that clearly consumed a quantity of beer. Part of the reason for this was the variable quality of drinking water, and it may have been the case that Selkirk ale was not strong. But its consumption was clearly bankrupting Thomas and his behaviour was exasperating his brothers. There is no record of the beer-swilling lawyer after this date, but John did go on to set up a legal practice and kept a Protocol Book between 1629 and 1633.

20 June

My neighbour has begun to cut the lush grass parks on either side of the Long Track to make hay. The air is heavy with the scent of all that green goodness.

21 June

This is the day of midsummer, the longest, the time when the sun climbs to its zenith and when in place of night there is the summer dim. In the small hours there was enough moonlight to see that the livery horses in the Tile Field were lying down. Flight animals, one often remains standing guard, watching for predators in the shadows. Horses always choose open ground to lie down, but how they decide which of their number should watch over them is mysterious, secret. The American author Jane Smiley once wrote that in a horse's eye there are things beyond comprehension.

Across the valley, green barley fields ripple in the breeze as Maidie and I walk out. Flowers carpet the grass parks: daisies, their tiny white petals tinged with crimson, buttercups egg-yolk yellow, and the white crowns of clover are everywhere. Yesterday Grace and I counted eleven sorts of wildflowers around her house, and one pansy that had self-seeded from her grannie's pots. The wee one hunkers down to pick the daisies and presents them as gifts. My pockets are full of little dried-up blossoms and one sits at my elbow as I write this.

When the sun begins to die in the west late tonight, I shall fire up my quad bike and go to the Bronze Age fort that dominates our valley so that I can watch it slip behind the hills of the Ettrick Forest. Perhaps ghosts will flit amongst the shadows of the trees, whispering of ten thousand midsummer eves.

22 June

From the summit of Soutra Hill, the watershed ridge between the Lothians and the Forth and the Tweed Valley, I gazed over the majesty of Creation. The midnight light in the north glowed dusky yellow as the unset sun moved slowly behind the mountains, just below the horizon, edging around the rim of the world to meet the morning. The vastness of the cloudless, pale-blue Heavens soared above me and dimmed towards the southern darkness.

I was driving home from a book launch at Dunfermline Abbey. My old friend Gordon Brown and I have written a history of Fife, and four hundred people filled the old church to listen to us talk of history, of kings, saints, miners, weavers, Andrew Carnegie, North Sea oil and the uncertainties of the future. The abbey is a thin place where the veil between a long past and the present is no more than gossamer. Beneath the flagstones where I stood, Alexander III of Scotland is buried, his brains dashed and limbs mangled after a fall from his horse down Fife's Pettycur cliffs on a stormy night in 1286. His death plunged Scotland into two centuries of war with England. Behind me was the tomb of the man who eventually succeeded Alexander, the saviour of Scotland, the victor of the battle for a nation at Bannockburn, Robert Bruce. And beyond the apse stand the remains of the shrine of Holy Margaret. A Saxon princess married to Malcolm III Canmore, a roaring, Gaelic-speaking, bearded king, she tamed the wildness of his warrior court and her piety earned her sainthood and enduring reverence. It was a privilege to speak there and listen for the long, faint echoes of Scotland's past.

The magnificent reach of the three Forth Bridges are our versions of the great churches, abbeys and cathedrals. The new Queensferry Crossing is elegant, monumental, like the flying buttresses at Dunfermline and the shattered spires of the ruined cathedral at St Andrews. Driving home, I had moved seamlessly between ages, between worlds, and to stop, stand on Soutra Hill and look back to the undying midnight sun, I thought of continuity in this cradle of Scotland, the lands on either shore of the Forth, of the generations that had passed into darkness and those that are to come. Up on the windy hill, perhaps for a fleeting moment, I could touch the edging light of eternity.

Warmed by the soft sun of the morning, Maidie and I walked along the Top Track. I wondered if Walter Scott talked to his dog as much as I do. I hope so. Rounding the corner at Windy Gates, I saw a stoat skipping over the rows of cut hay-grass before disappearing into the safety of the hedge. The sweet scent of cut grass was already heavy in the air.

23 June

Until today we had no idea if the house martins were coming or going. At the end of May four arrived after the long flight from Africa and began to refurbish last year's mud nests under the eaves above the porch. Ruthless colonisers, sparrows tried to take over these ready-made pods; to discourage them, we suspended a row of shiny washers on strings from the guttering. This contraption was supposed to discourage the sparrows and allow the more athletic martins to fly behind the dangling obstruction to reclaim their property. It did not work. Now we have a noisy, squabbling extended family of sparrows and no martins.

But this morning I saw six flying at great speed around the farmhouse, the white stripe on their backs unmistakable. They swooped down to where the trickle from a broken stone drain makes mud on the side of the track and were scooping it up in their beaks. Nest-building! Better late than never. But where? Martins will not build nests on wooden buildings because the mud pellets do not stick to the smooth, painted surfaces. Only stone will do, and the only stone building is the old farmhouse. And I can find no sign of anything being built under the eaves or on the walls. Mysterious martins. But at least they are here and we have the daily joy of watching their aerobatics.

My neighbour has cut about half of the lush grass in the parks on either side of the Long Track. Starting by skirting the margins, he has cut perhaps ten unbroken rows, looping the field in lazy oblongs, leaving a large island uncut in the middle. It looked as though God had been drawing with a green felt tip pen on the land. Last night my neighbour picked up the rows and round-baled them. Some rolled downhill into our boundary fence.

The effect of this harvest is very beautiful. Where the cutter has been, its passes have shown up the sinuous, almost voluptuous contours of the land, its lines describing long S-shapes through the swales and shallow hollows. Between the pale, lime-yellow of the cut margins and the dense green of the grass still standing, there is an eye-widening, eye-filling contrast. In their stunning works of land art, Charles Jencks and Andy Goldsworthy see and accentuate these natural forms and processes. And are much admired by art critics and the public. My neighbour does it every summer.

The morning was so still not even the spindly stalks of cow parsley were moving in the verges of the Long Track. Looking at the lush landscape, I lost all sense of time and stayed out for much longer than my usual forty minutes with Maidie.

At 5 a.m. tomorrow I will leave on an expedition. To research a book on the Christian conversion of the Hebrides, I will be going island-hopping. Columba is only the most famous of these leathery old saints who founded monasteries and hermitages down the Atlantic shore in the sixth and seventh centuries. My first landfall will be on Eileach an Naoimh, a rocky, uninhabited island where the wide mouth of the Firth of Lorne opens onto the mighty ocean. Islands and the open sea make me nervous. I fear the deeps because I cannot swim and worry about being stranded on an island in a storm. No mobile phone signal to summon help, no shelter except rocky overhangs, and only Co-op pork pies and Cadbury's Fruit & Nut chocolate standing between me and extinction. But how else can I write about these men and their mission unless I sail where they sailed and walk where they walked? I have packed, repacked and checked the weather forecast every hour. Thunder and lightning overnight and rain clearing in the morning.

It will be an adventure, a journey to a different Scotland, and not part of this story. Unless I fail to return . . . in which case this will be the last entry . . .

28 June

I have returned, more or less in one piece. Walking out with Maidie in the softness of the morning mist, I felt myself exhale after four frantic days of ferries, island-hopping, map-reading, note-taking and hundreds of phone photographs. After the drama of the Hebridean landscape, I felt a peace dripping slowly from the clearing skies and knew that the sun would make the day glow in this quiet, fertile and familiar landscape. It was good to be home, good to resume the unthinking routines of feeding dogs, of leading out horses and of sitting at my desk scribbling. I see an email from Rory Low about some new Elizabethan finds and I will try to find a day for a systematic survey of the Doocot Field with Walter Elliot and his magical divining rods. I am certain there is a story waiting under the grass.

29 June

Out early in the summer warmth, Maidie and I were startled by a sharp squawk in the skies above the Top Track. Like a squadron of spitfires, six herons were flying south towards the Tile Field. Their grey undercarriages and their tucked-in necks were clearly visible. Herons are usually silent, even dignified, aloof, as they glide over us on a higher plane. I wondered and wished I understood more about the calls of birds, what they are communicating and why those solitaries were flying in a noisy formation.

In hedges at the edge of woods and intertwined with shrubs, the dog roses have come out all at once, it seems. Five-petalled, milky-white tinged with pink and perfectly symmetrical, they looked heraldic, like the Tudor rose, a diplomatic combination of the red of the Lancastrians and the white of the Yorkists. Their scent was delicate and all the sweeter for that.

30 June

The Common Burn has completely dried up. An ancient boundary, the resort of frogs, even the very occasional otter, it no longer flows through our history. Despite frequent and prolonged wet weather, and the surprising fact that it still ran after the long, dry spell of six weeks last May and June, there is now nothing but a bed of damp, fissured mud. Somewhere along its length, something has changed.

Now that my neighbour has harvested his first cut of silage, completing his round-baling in yesterday's warmth, I can walk over the grass park between the Long Track and the course of the burn. There is an old Irish bridge, a wide pipe long ago laid on the bed of the burn and then covered with earth, that links the park with the Tile Field and I wondered if it was blocked. But there was nothing. And both up and downstream no water ran. Tall grasses grew where once there was a flow. Plenty of water runs under the stone bridge that carries the Long Track to the Old Roman road and its modern tarmac. So there must be a blockage between there and the Irish bridge. I will walk that line as soon as I can.

If the water table has dropped dramatically and suddenly, it will not only make frogs and other aquatics homeless and cease to be a corridor for otters (there is fencing on either side of the burn to create this), it would also not be without precedent around our farm. Such changes in the past have had serial and even calamitous effects. Selkirk's burgh records note a bitterly fought legal case that revolved around the careless management of water. In 1661, John Riddell, Laird of the Haining, drained the loch in front of his house, lowering its levels by running it off through a remarkably named stream. The Clock Sorrow runs behind the Riddells' mansion and downhill to the River Ettrick. The first element of the name derives from the Latin *cloaca* for a sewer or a drain. In Ancient Rome the Cloaca Maxima ran into the Tiber, carrying what the citizens called *acqua nera*, black water. What caused lawyers to intervene in 1661 was brown water.

Beneath the Haining Loch several ferrous springs bubble up and sometimes make the waters of the loch look very turgid. Pike prosper in this highly mineralised mixture, but other species have little or no tolerance for iron. When Riddell drained the water through the Clock Sorrow and into the River Ettrick, it had such a catastrophic effect on the famous salmon fisheries at the mouth of the Tweed that the Mayor of Berwick instructed lawyers to issue a writ for damages. The Ettrick is a major tributary of the Tweed, and as the iron-rich water clouded the river it killed the salmon. The second element of the name of the Clock Sorrow might be obscure but it was apposite in 1661.

Maidie was unimpressed by my search through the dense grasses, thistle and nettles on either side of the dried-up Common Burn. When DEFCON 1 ignites, she will launch herself ferociously at anything with four or even two legs, regardless of their size, but the little terrier refuses to go through nettles or thistles despite her thick, white coat. She simply sits down, sets her jaw, stares at me and refuses to budge until I sigh, walk back and pick her up. I am glad that at 7 a.m. no one witnesses this.

July

1 July

Recently I came across a remarkable document, another crack in the continuity of time that stories can slip through. Three hundred and forty-eight years ago, a stud for the breeding of thoroughbred horses was almost set up on our farm. A contract was agreed between John Riddell of the Haining, the arch polluter, and King Charles II. The whole estate was laid out for breeding. Stables were built (their successors still exist), paddocks fenced and all the necessary arrangements were made. It seems that Charles II had agreed to send a stallion to the Haining, but he never arrived and the king was forced to compensate Riddell.

In 1603 King James I and VI visited Newmarket Heath in Suffolk and pronounced it perfect for horse racing, a sport he loved. His grandson was equally besotted, and Charles II even rode as a jockey at Newmarket. He always won. At the end of the seventeenth century what are known as the foundation stallions began to arrive in Britain. These were much admired Arab horses from the Middle East, difficult to acquire, famed for their fineness, stamina, intelligence and speed. The Byerley Turk was captured from the Ottomans by Captain Robert Byerley at the Battle of Buda[pest] in 1686. He then charged in the van of King William of Orange's cavalry at the Battle of the Boyne in 1690 before standing at stud (the horse, not the captain) at Middridge Hall in County Durham. The Darley Arabian was bought in 1704 by Thomas Darley, the British Consul at Aleppo in Syria, for the fabulous sum of 300 gold sovereigns, and he stood at Aldby Hall in Yorkshire. And the final foundation stallion, the Godolphin Arabian, was born in 1724 and bought by Edward Coke of Longland Hall in Derbyshire. All so-called 'blood' horses are the descendants of these three Arabians, including our own beautiful black filly, Sula's Imprint. That means Topsy, our thoroughbred mare, has at last fulfilled the royal contract, three and a half centuries after it was agreed.

2 July

The rabbit population has exploded. The lush, hot and damp summer has supplied juicy grass and every morning they scutter everywhere, even a Beatrix Potter-perfect baby hops around the terrace, driving Maidie into paroxysms of rage. When I finally managed to drag her out this morning, the cows at Huppanova were lying down with their calves, already sunbathing, and as we passed the shade of the Haining Wood I noticed that the trees seemed to make the birdsong echo.

3 July

Dragging a reluctant Maidie behind me, I walked the line of the Common Burn to where I thought it spurred off from the Hartwoodburn. There was no flow at all, just a damp seed bed for tall, dew-soaked grass. When I reached the junction with the Hartwoodburn, I had to climb the barbed wire fence of the otter corridor and thrash my way through the nettles and willowherb to get close enough to see if there was a blockage. Cocking her head quizzically from side to side, Maidie listened to me curse as I was stung and pricked.

There was no blockage I could see, just a six-inch sill carved out by the Hartwoodburn as it turned into the shadows of a sitka wood. Except in winter, I suspect, when rain was heavy and persistent, the Common Burn has never taken its water from the Hartwoodburn. The sill was stony and compacted, well established and definitely not a pile-up of silty mud. All of which led me to wonder where the burn's flow of water had come from and why it had suddenly dried up. The grass park next to the Long Track slopes towards it and rain run-off will have seeped in, and there is also an old spring that rises in one of the swales of the park, near our boundary hedge. Perhaps that has failed or its water has found a different course. The only obvious recent change was my neighbour's repair of his drains in the Tile Field, but there is a slight hollow where he dug and no obvious fall towards the Common Burn.

Soaked up to the knees but happy to walk and dry off in the warm sunshine, I saw sheep moving across the southern ridge and could hear the gentle purr of the shepherd's quad bike. All was unhurried as his dogs loped uphill, circled behind the flock and began to push them towards the gate to fresh pasture.

4 July

The old spring from the grass park that trickled through the wood behind the stables to feed the Common Burn has also completely dried up. Arms in the air, wading through chest-high nettles and tangles of sticky willy, I could see nothing but a dark bed of mud. No water had flowed there for some time. Only a rainwater drain was dribbling into the burn. Something radical has changed. If the water table has dropped, then other burns should be affected. But the Nameless Burn in the East Meadow and the Hartwoodburn are still running. Their levels are what you would expect in July, but the water is clear and moving. I am vexed at the loss of the Common Burn and its habitat, and even more vexed that I don't understand why it has gone.

5 July

When Rory began to find many medieval coins in the Doocot Field, he was careful to mark the find-spots accurately. This mosaic of dots greatly intrigued Walter Elliot, and me, because it was most dense on either side of the Long Track and at the northern end of the field, beyond the ruins of the doocot, not far from the site of Selkirk Castle. Yesterday afternoon, in spectacular fashion, Walter joined up the dots.

After decades in the fencing business – when he used divining rods to discover hidden ditches and postholes, thereby saving a lot of unnecessary digging – Walter turned his techniques to archaeology. No one, including Walter, understands exactly why it works and how two metal rods bent at right angles and held loosely in the fists suddenly turn inwards when a feature is detected. But it does work. Not everyone has the ability to divine, and when Walter taught me how to hold and use the rods many years ago (we were looking for the course of a water pipe on the farm) I had results immediately, whereas when Lindsay held the rods, there was no movement.

Now eighty-four, Walter has been very unwell of late and it has taken a few weeks for him to recover. But when we drove up the Long Track into the Doocot Field, he jumped smartly out of the pick-up and began looking around. Walter has a well-honed eye for the lie of the land and, handing me a spare pair of rods, we walked up the side of the Long Track. As we moved slowly, staying abreast, about five yards apart, both sets of rods immediately began to swing at exactly the same moment. Pacing out from where they first moved, Walter saw the rods swing again and soon he had worked out that we were standing on a rectilinear, subterranean feature measuring about five metres by three metres and with a door on the track side. Almost certainly built of wood, it was probably a cottage or perhaps a workshop of some kind. Then we found another, and another, a line of three buildings running by the side of the track. Between two of them was space for a lane.

In less than an hour we had discovered what is almost certainly the faded outline, no more than a distant echo, of an early medieval village whose buildings had long ago disintegrated and sunk back into the soil. But their wooden posts, and wattle and daub walls, had left a whisper of people, their conversations, their troubles and joys, the transit of lives lived perhaps a thousand years ago. There were houses on both sides of the Long Track and then another row running at right angles, towards the site of Selkirk Castle, a pit behind one of them, and the remains of their backlands, the fenced areas where they often kept a cow overnight, where they dumped rubbish and sewage and sometimes grew vegetables. Beyond these we found the long ago ploughed-out ditches of runrig cultivation and they drained downhill towards the Haining Loch. Under the lush summer grass lay the watermark of a lost village. It was absolutely exhilarating to spend the afternoon searching with the rods, listening to Walter work out what we were finding, bringing all of his vast experience to bear. The site was completely unsuspected until Rory began to find coins and other items, and what Walter and I discovered brought the otherwise ordinary field to vivid life.

Rory's finds, not only of coins but also of weighing apparatus, the large piece of lead and much else, appear to have been the sole surviving deposits, in metal, of a community that lived in wooden houses, organic material that has completely disappeared. Walter wondered if this was the early and original Anglian village of Selescirce, 'the church by the hall', that eventually gave its name to the town of Selkirk. It was first recorded as Seleschirche in 1124, soon after the castle was built. Its location at the north end of the Doocot Field also places it near a road junction, where the Long Track joined with the road from the valleys and the west, where I had walked one evening a few weeks ago. A faded map of the long past seemed slowly to be coming back into focus. There are certainly more stories lying hidden under the grass where the cows and their calves browse and where I had walked often with Maidie. But they are no longer lost and forgotten, these people who once lived where we live now.

6 July

Having shed their tiny white flowers, the wild raspberries are beginning to form on the canes that line the Bottom Track. When ripe, they are sweet and tart at the same time, and the trick in picking them is timing. When the berries turn red and become easily visible, the birds peck at them and seem not to care about ripeness. I hope to harvest what they leave at exactly the right moment. Wild rasps and Greek yoghurt make a good summer breakfast.

This morning I was surprised to see the lime-green of a cut grass park by the Thief Road, uphill from Hartwoodmyres, where the Ettrick Hills begin to rise in the west. Never having seen that before, I later looked on the Pathfinder map to see what the altitude was. Harvesting a grass crop, either hay or silage, at nine hundred feet is new, at least to me. We lie on the six hundred-foot contour. Perhaps this is another sign of climate change. When the vacuum that is political leadership across the world fails to deal with this emergency, as it surely must, and the processes become irreversible, the great migrations will begin. From the increasingly uninhabitable lands around the equator, and even southern Europe, people will be forced to move north. If the warmer and wetter climate expands, the land available for cultivation, even if only for grass, in Scotland we will become an attractive destination. Lying at the farthest north-west end of Europe, and with wide areas very lightly populated, we should expect to see more and more people come, many of them desperate.

8 July

The salt sea air of the islands on the edge of the ocean startled me when I reached the west coast two weeks ago. Thinking of the stink of fish and diesel swirling around the quaysides at Oban, the spray on my face as the ferry carried us across the Firth of Lorn to Lismore, brought back to me the absolute dominance of the sea and how it overwhelms the senses of a landlocked person like myself; sight, smell and sound are all very different. Our farm lies almost fifty miles from the North Sea in the east, another fifty from Edinburgh and the Firth of Forth in the north, and more than fifty from the Solway Firth and the Irish Sea. We are a long way inland. Here the air is earthy, sometimes even scented, sounds are not echoic and the hundred colours of green stretch away on every side. All very different – so familiar that I find the details of that difference difficult to describe. It is the air that I breathed when I was born and when I roamed the countryside as a boy and a young man. And I shall breathe my last here.

Coming back home to live in the Borders was more of a reflex than a conscious choice, but it is good to be refreshed by other worlds, different climates and cultures. Until I first travelled west to the Highlands and Islands as a teenager, I had always associated the seaside with ice cream, candy floss, sandy sandwiches and amusement arcades. Each summer the churches in Kelso organised Spittal Trip for the children who attended their Sunday Schools. Spittal is a small seaside resort with a long sandy beach near Berwick-upon-Tweed. In the weeks prior to the trip, attendance at Sunday School was never less than perfect.

This excursion bore all the lineaments of Victorian charity – a day out at the seaside for children too poor to have holidays – but it was wonderful, the exuberant joy of an old-fashioned excursion rising above Presbyterian propriety. Each church's Sunday School gathered early in the morning of the appointed day and cardboard lapel badges were handed out, as if we were wartime evacuees. Packets of coloured streamers were issued to older children and we marched uphill to Kelso station, where the special train waited, puffing and huffing before it clanked eastwards. We all cheered as it picked up speed. The small top windows were slid open and the streamers jammed in. Building anticipation, competing to be the first to see the sea, we cheered again as we passed every village station: Sprouston, Carham, Cornhill, Twizel (where a spectacular and scary viaduct crossed the River Till), Middle Ord, East Ord and finally Tweedmouth station. Marshalled once more in military order, we marched first to Spittal's church halls, where we lined up to be given brown paper bags of buns, a chocolate biscuit and sandwiches, as well as a split of Middlemas' lemonade (plain). And then we marched down to the beach. Dog-collared, black-clad ministers and kirk elders ensured that there was some decorum and not too much shrieking when children splashed into the chill North Sea.

I have a photograph of myself with my sisters and Ronnie and Suzie Taylor sitting amongst the deckchairs, all wearing overcoats and cardigans, lapel badges dangling. It is sunny, but it must have been a cold day. Every face is wreathed in smiles. Charity certainly, but very cheerful.

9 July

Poison lines the Long Track. Some of what I had previously believed to be cow parsley turns out to be hemlock, what Socrates used to commit suicide. This morning I took my Collins Gem guide to wild flowers with the intention of identifying a clump of very pretty white flowers by the side of the Top Track (stitch-wort, not sure if Greater or Lesser). When Maidie and I turned down the Long Track, I looked up the cow parsley family (they are basically carrots) and, flicking through the pages, I recognised hemlock growing by the fences of the grass parks. It has feathery leaves and a purple stem, and it smells bad.

After early morning rain, scents swam in the warm air: honeysuckle by the Wood Barn, pine everywhere, the bitter odour of new leaves in the hedgerows and, because there is so much of it, the faint smell of elder blossom drifted over the track into the Deer Park. The thistles were beginning to put out their purple crowns. They have a sweet smell and, more surprisingly, they are part of the daisy family, something that does not quite chime with the thistle's spiky role as an emblem of spiky Scotland.

10 July

Yesterday evening an aerial battle raged in the skies above the stables. Because their first broods of chicks have hatched, squadrons of swallows were sallying out to see off crows who had approached too close. Using their extraordinary speed and agility, these little birds buzz the black crows like spitfires attacking bombers. Another patrol saw off a kestrel so that the no-fly zone was not breached.

The day dawned with heavy rain, giving way to a close, muggy, moist air, the perfect atmosphere for midges, tics, clegs and the myriad biting insects that thrive in these conditions. Outside the window of my office is a cotoneaster bush in full flower and the drone of a hundred wasps is audible as I write this. There seem to be more bees this summer, and we must learn to love not only them but all our pollinators, whether or not they bite or sting.

11 July

The loud cracking and whumping became immediately audible at Windy Gates. A tree harvester, a vicious, predatory monster of a machine, had cut a wide swathe through the wood around the Haining Loch and trees were being felled at industrial speed. Many decades of growth brought crashing down in seconds. In

2009 the Haining House and stables, 160 acres of woodland around it, and the loch were gifted in his will by Andrew Nimmo-Smith, a descendant of the Pringle family who came here at the beginning of the eighteenth century, to the people of Selkirk and Selkirkshire. The policies had not been well maintained and the wood around the loch certainly needed to be thinned, but the clear-felling is a savage process. The sound of the harvester spooks our horses, making them hard to catch in the evening.

I hope they leave the hardwoods on the margins of the wood, especially the stand of scented sweet poplars in the corner by the Deer Park track.

12 July

By the end of yesterday afternoon, many of the trees at the southern end of the Haining Loch that used to border our land had gone. This morning the effect was very dramatic, opening up vistas that had been closed for many decades, revealing much that had been hidden. The cutting of healthy trees always vexes me, but, after standing for a long time looking at the nature of the changes, I decided that it was for the best.

For many years the woods had not been managed and were choked with self-seeded trees. Most of these were scruffy sitka spruce, many of them spindly, as their tops struggled to reach up through the tangle to the light. They are certainly not an adornment, but I was sad to see that the stand of sweet poplars had gone and I shall miss their metallic scent next spring. Woodpeckers often drummed on the bark of older, half-fallen hardwoods in that corner and I am sad that they too have been cut. These busy little black and white birds will need to seek new stores of insects. As Maidie and I surveyed the revelations of the cut wood in the sort of silence that might have followed a battle, I saw a stoat darting in and out of the logs, working out the radically new geography of his territory.

Once it is complete, the cut logs removed and the brash burned, I think the felling will open up a new landscape and show off its gentle beauty better. Dense woods of sitka tend to close down the shape of the ground, blanking off wide areas, burying them in sterile darkness, making them impenetrable. But now the ancient bounds of the Deer Park are much clearer, with the enclosing bank and perhaps a deer-leap revealed, if it is not ripped to pieces by the caterpillar tracks of the harvester. I shall have a closer look when the monster machines depart.

We can now see from Windy Gates how all of the trees we planted on the western slopes of the Deer Park are growing, what sort of pattern they will make when mature. The loch is now clearly visible rather than just a glint through the leafless trees in winter, and at its far end we can see the foursquare shape of the Haining, its Georgian symmetry very pleasing against the fluid lines of the landscape. It looks like a view of a new aristocratic mansion commissioned from Canaletto. The pellucid waters of the loch set off the upright mass of the house and the trees look as though they have been artfully arranged to frame it. Next to it stands a marquee, no doubt much used for weddings, a booming trade. Fooled by their satnav, or perhaps just fools, several sets of guests have driven down our track looking for champagne and canapés, thinking that our humble farmhouse is the Haining.

Just as Maidie and I turned for home, the morning sun lit the cut wood, its rays splashing yellow over the piles of logs and the revealed ground. It will have been many, many years since such a brilliant light warmed the sterile slopes where the sitka grew.

13 July

More poison is growing along the margins of the Top Track. Several small stands of ragwort, highly toxic for horses, will have to be dug up before it flowers. Because the sap from the stems causes a very angry skin rash, gloves will be needed. The warmth and wet of this July are creating a temperate jungle where everything is flourishing, bad and good, and I am relieved to be irritated, at last, by the flies. They are everywhere and we have never seen so many bees. Yesterday evening we sat outside on the terrace, where Lindsay has planted many lovely little shrubs amongst the paving. One is particularly beloved by the bees and perhaps twenty were harvesting the pollen, tiny lumps of it sticking to their legs. Driving up the Long Track puts up more butterflies from the verges than last year, particularly red admirals. Perhaps fewer species of insects are dying than is predicted.

Yesterday afternoon a huge truck came barrelling down our track, stopping outside the farmhouse, the driver thinking he was going somewhere else. Misled, yet again, by satnav, he had great trouble turning his belching leviathan and he rutted the tracks as his vast tyres ripped through the gravel. Action was overdue and I drove immediately to Gala to have large signs made. They will be planted prominently at Windy Gates and, leaving aside the precise wording, the basic message will be 'BUGGER OFF'.

15 July

Early on an autumn morning in a quiet meadow east of Selkirk, two men faced each other, their swords drawn, their tempers flaring. A duel had been sparked by the greatest, most weighty political issue in the nation's history, one that drove deep disagreement into passionate hatred. One that still simmers in the twenty-first century. Mark Pringle of the Haining, the youngest brother of the laird, John Pringle, believed absolutely that the Union of the Parliaments of Scotland and England and Wales was not only in the best interests of both countries but also of his family. Glaring at him across the dew-soaked grass was Walter Scott, Laird of Raeburn, and a distant ancestor of Sir Walter

Scott, the great novelist and poet. He had been raised by another Walter Scott, known as Beardie because he refused to shave until the Stuarts were restored to the throne of Great Britain and Ireland.

On the evening before the duellists met in the meadow there had been a meeting in Selkirk to discuss the business and government of the county. On 2 October 1707, lairds, burgesses and other men of importance had ridden to the county town to make their views known, probably to John Pringle of the Haining. He had been elected MP for Selkirkshire to the old Scottish Parliament in 1703 and had been transferred to the new one at Westminster in 1707, one of only forty-five Scottish representatives. The parliament of the United Kingdom had convened on 1 May and it may be that Pringle planned to say something of its workings, the opportunities and perhaps the difficulties. No minutes were recorded for the meeting of 2 October, but another similar gathering was arranged for 28 October and a record of what was discussed has survived in the Walter Mason Papers. Money, the amount and rate of taxation for the county to be collected, was the main subject for discussion and decision, as were the excise duties to be levied. Matters were complicated by the different values of the pounds Scots and the English, or pound sterling.

After the business was concluded on 2 October, the lairds and the other notables sat down to dinner in one of Selkirk's inns and a session of heavy drinking followed. As the ale and claret flowed, a furious disagreement erupted between Walter Scott and Mark Pringle. Very few details survive, most of them little more than much-repeated and embellished rumour, but it seems that both parties eventually became so drunk that tempers cooled. In fact, all the sources agree that neither man could remember anything about the quarrel, except that there had been a quarrel.

Early on the following morning a peculiar, arcane ritual was enacted. No doubt hungover, bleary eyed and tired, Scott awoke to discover something that sounds very odd, that he had 'bitten his glove'. This signified that some deadly insult had been thrown at him by Pringle as bottles and tankards were drained after dinner. Even though Scott could not remember what the insult was, it needed to be avenged; he had to have satisfaction. The business of the bitten glove is explained in Sir Walter Scott's *The Bride of Lammermoor*. A variation on throwing down the gauntlet or slapping an opponent's face with a glove, it meant that Scott of Raeburn had pulled off his glove with his teeth. That somehow proved that a deep offence, which no one could remember, had been flung at him by Mark Pringle, and only blood could avenge it.

To add even more confusion, other versions of the story have Pringle recognising that his glove had been bitten and then galloping from the Haining to Galashiels to call out Scott from the house of his brother-in-law. However all that may be, two young men, accompanied by their seconds and probably some others who had been at the fateful dinner, confronted each other in the meadow. As they drew their swords, it seemed that the past was confronting the future. A supporter of the Stuarts, of a return to Catholicism, of a much more absolute monarchy, was about to contend with a man who believed in limiting the power of kings and in the good sense of uniting the parliaments of Scotland with that of England and Wales.

The duel began with flurries of swordplay, both men apparently being experts with the small sword, a sort of rapier with a very sharp tip. At first only steel clashed and no blood was drawn. After a time Mark Pringle called for his opponent to desist, reminding Scott that he was married with three young children and, since honour had surely been served, what point was there in continuing a dangerous contest. Taking that as an insult, Scott would have none of it and the fighting intensified. And then, in a moment, it ended. Having parried a stroke and thereby brought his opponent's guard down, Pringle lunged forward and drove his small sword through Scott's body. Those watching gasped. Drawn blood from a cut would have been a sufficient sign of victory, but death was a shocking, unlooked-for outcome.

As Scott lay on the grass, bleeding to death, Pringle fled first to the Haining and then rode hard to Leith, where he took passage on a ship bound for Spain. The Borders was scandalised at Walter Scott's killing, no matter the circumstances or the pleas to desist, and criminal charges would undoubtedly have been brought. Exiled for more than thirty years, Mark Pringle eventually followed the example of many younger sons of landed families. He set up as a merchant (perhaps with the help of his elder brother) and in 1738 returned to Scotland a wealthy man. Old scores were certainly not settled, for no prosecution was brought, but Pringle had the good sense not to come back to the Borders. His fortune enabled the purchase of the Crichton estate in Midlothian.

In 1765 a complex legal wrangle riddled with contradictions, claims and counter-claims began. When Mark Pringle died, he left the bulk of his estate to his eldest son, John, but following a deathbed instruction he attempted to settle £1,000 on his youngest son, Mark. John objected to this and the case dragged on through the Court of Session for some considerable time. Then fate and fortune took another twist. Through the accidents of inheritance, a lack of direct heirs and a premature death, Mark Pringle (jr) took possession of the Haining estate in 1792. With the eager help of architects, he began to think about how a new and grander house could replace the old mansion at the head of the loch. And it seems that he also had a smaller, but apparently very immodest, house built.

When Lindsay and I first came to look at the cottage we later enlarged into the present farmhouse, we noticed a heavily weathered sandstone plaque above the lintel of the front door. It read *Marc. V. Pringle 1821*. And so it would seem that the son of the murderous duellist had our house built. Or did he? And why? As often happens, one story pulls aside a curtain to reveal others, some of them very unexpected.

16 July

Talking this morning with the foresters who are cutting down the wood around the Haining Loch, I was pleased to hear that the badgers had been successful in their parliamentary lobbying. Legislation to protect them has come into force. A colony of these little grey bears has dug a labyrinthine sett on the edge of the wood and under the fence that is our boundary. Exposed by the removal of the trees around it, the scale of the sett is much larger than I thought, about thirty yards by ten, and the earth excavated by the powerful claws of these creatures had mounded up above the fence posts. Because their breeding season began in late February and their cubs are beginning to emerge, the foresters have ringed the sett with a twenty-metre exclusion perimeter. The huge harvester cannot work there because its weight would have crushed the tunnels and sleeping chambers, so the trees will have to be cut with a chainsaw. What the badgers will do when they see that the comforting cover of the night wood has disappeared, I don't know.

18 July

Tabitha Twitchit is coming. Rather than stepping out of the pages of Beatrix Potter, a real tabby cat is coming to deal with the rapidly rising rabbit and mouse population. Unlike the original, we hope this cat will not be mumsy and kind but savage, a hunter, or at least one with a strong scent that will discourage the multi-plying bunnies and mice. Her prowling around the house will send Maidie into transports of rage, but not for a while. Tabitha is still a suckling kitten, the epitome of cute, and when she is weaned in a few weeks' time she will live at first in one of the empty stables.

19 July

When Andrew Pringle bought the Haining estate in 1701 for his son, John, he made him more than a landowner. In order to become an MP, a substantial property qualification was necessary. And to make the investment worthwhile, great opportunities appeared to lie just over the political horizon. John was duly elected in 1703 as the member for Selkirkshire.

After a series of severe crop failures and near-famine in the 1690s, and the catastrophic failure of the Darien scheme, Scotland's attempt to found a colony in Panama, the pressure for parliamentary union with England, Wales and Ireland intensified. Most Scots were against it, and it became clear that if the merger went ahead there would be no general election for the new parliament. It was deemed too risky to allow even the tiny minority who could vote to express their preferences, and during any election the much-feared Edinburgh mob would take to the streets. Instead some MPs already sitting in the Scottish Parliament would simply move to Westminster, as John Pringle had.

When all of this duly played out and Pringle took his seat, the MP was given the lucrative office of Keeper of the Signet in 1711 and also prospered in other ways. To overcome objections to the principle of union, a large cash sum was paid to Scotland, in effect to those MPs and others who ran Scotland. Known as the Equivalent, it was described as compensation for taking on a proportion of England's national debt. In reality, it was a large bribe that was distributed amongst the elite. Perhaps the Equivalent and the benefits to the Pringles of sitting in the Westminster parliament sharpened Walter Scott of Raeburn's resentments.

In any event, this new source of wealth allowed John Pringle to begin to change the landscape of the Haining and of our farm.

20 July

As a rosy future unfolded for the occupants of the big house at the Haining, their much more humble near neighbours were slowly sinking into a grassy oblivion. The little village Walter, Rory and I discovered at the north end of the Doocot Field had probably completely disappeared by 1700. Yesterday evening Rory was out again with his metal detector. By the side of the Long Track, where the divining rods had been swinging, more Edward I silver pennies came out of the ground. There are likely many more because Rory is finding the lush grass of this humid and damp period very troublesome. The cattle simply can't eat it down fast enough and its density is muffling the signal of metal buried at least a foot deep. Rory would love to scrape off the top six inches of soil; he is certain that much more of the story of the little medieval village would then be revealed.

The chronological clustering of the coins is very suggestive of what happened to these wooden houses and workshops. More than thirty silver pennies and other coins have come up, most of them later thirteenth century, many with Edward I's head on them, some from very specific English mints. But very little can be dated to the fourteenth century and nothing at all is later than that. It seems that the hamlet astride the Long Track thrived on trade, but after the beginning of the wars of independence and the centuries of cross-border warfare, commerce was severely disrupted.

Twenty miles to the east, the town of Roxburgh stood on the haughland opposite Kelso, where the Teviot joins the Tweed. It was where urban life began in Scotland, the first town of any size, and from the early twelfth century onwards a brisk trade in wool and hides was conducted at its markets. Much was despatched down the Tweed, possibly by raft, and exported from the quays at Berwick to the textile manufacturers of Flanders and northern Italy. Business boomed, the town expanded and local producers prospered. But when Edward I's armies began to march north, trade was badly disrupted and never recovered. By the sixteenth century, this thriving town, with its four churches, a royal mint and a grammar school, had almost completely disappeared. Now there is nothing to see, not one stone left standing upon another, just a wide and lush grass park where cattle graze. On a much smaller scale, a similar decline took place in the Doocot Field.

21 July

It is good to be old and growing. Yesterday evening I dug more of our new potatoes, an excellent crop this year. Steamed whole, split open for butter and sprinkled with salt and pepper, they are a meal in themselves. Sweet, clean-skinned and surprisingly large, they grow in three raised beds in the small paddock behind the stables. Filled with well-rotted horse muck (we have an endless supply) mixed with soil finely tilthed by the moles as they dig their winter tunnels, they are beds of great fertility. The rain and the warmth helped produce such lush shaws that I thought they might have used up all of the plants' energy and the potatoes would be small and sparse. But there were plenty, some the size of oranges. From my conservatory/greenhouse, I pinched out enough lettuce and picked several small tomatoes to make a salad to go with the spuds. We didn't really need the sausages (made by the excellent butcher in Selkirk from local beef).

Home-grown food tastes good because, unlike what we buy, no chemicals or refrigeration are used to preserve it on its journey from the producer to the shops and then to the kitchen table, and because it is home-grown. The tilthed soil and horse muck is so powdery that I can dig the potatoes with my hands and do not have to risk spearing any with a fork. But not all my efforts have worked. Eight tomato plants have remained stubbornly barren and the rocket I sowed is refusing to grow, although its purple-tinged leaves look lovely. Next year I will have to rotate crops and grow carrots and spinach where the spuds are. That means I will need more raised beds for another crop of potatoes next summer.

22 July

Silent on a Sunday, the tree harvester and the forwarding trailers that take away the cut logs were tucked away in a corner of the wide area they have clear-felled in the Haining Wood. On a still evening, this scene of raw devastation has not only opened up the Canaletto view of the big house at the far end of the loch, it has also revealed the lost bounds of the medieval Deer Park. I had told the foresters where I thought the park pale might run (the unmanaged wood was so dense that I couldn't be sure) and they have left it intact where they could. It is fascinating to see the patient work of the deerherds revealed.

Impossible to date accurately (unless Rory finds a coin in one of the tumbled-down parts) but beautifully made, the pale has been revetted with stone on the inside and then filled and topped with earth. With the ditch dug deep below it, the revetted face would have made it impossible for deer to jump out of the park. Now saplings of ash, willow and other small trees grow out of the earth in the pale, but in the Middle Ages it is likely that quickset would have been planted. Also known as quickthorn, and nowadays as blackthorn, it grows very fast and its long, needle-like thorns make it impenetrable, even dangerous. Much cheaper, and just as effective as a stone wall, quickset was sown at the bottom of ditches around fortifications and it could be impossible to break through without serious injury. We made the great mistake of planting blackthorn in our hedges and it took me an afternoon of hard work, and a great deal of cursing, as well as many deep scratches, to cut down only four plants, clearing an area of about three feet. A waste of time.

Tabitha Twitchit comes tomorrow. Bunnies, watch out.

23 July

Buried under the long grass of the Doocot Field, Rory has found the remains of a love story. Out of the ground came an American Zippo-style cigarette lighter made in the shape of a miniature book and on the spine is clearly engraved *To Rita – Edek 1943*. Edek is the Polish version of Edward, and he was almost certainly a soldier with the Forward Artillery Regiment of the Free Polish Army. Thousands of Poles were stationed in Scotland during the Second World War and Haining House was requisitioned as accommodation for officers and as a headquarters. With many Borders men away from home, enlisted in the British Army, the dashing, well-mannered Polish soldiers were very attractive to local girls. My mother remembers them coming to country dances in their smart uniforms, asking very politely, clicking their heels, for a dance. 'And could they dance!' she said. 'Waltzes, polkas, everything!'

Rita must have been attracted, and when Edek gave her the engraved lighter, something she could show off to her friends, something more lasting might have been forming. After the end of the war and the Soviet takeover of their homeland, many Polish soldiers decided to stay on, marrying local girls. In my class at school was the first generation of Polish-Scots children: Wichary, Mazur, Tomczek, Goldstajn, Poloczek. Rory's find begged questions. How did it happen that Rita lost her lighter in the Doocot Field, by the Long Track? Perhaps on a summer evening walk from the Haining with Edek, having lit cigarettes, she simply dropped it, and despite frantic searching it could not be found. Or perhaps in a fit of disappointment, sad news or anger, she threw it away.

Billeted with the soldiers of the Forward Artillery Regiment was an unusual recruit. Formally enlisted as a private and later promoted to corporal, Wojtek the Bear lived for a while near our farm. Bought as a cub by the Polish II Corps when they were evacuated from the USSR to Iran, this Syrian brown bear became famous and was much loved. Fond of beer and coffee, he imitated what the soldiers did, even sleeping in their tents on cold nights.

After the invasion of Italy in the autumn of 1943, Polish units left the Haining to fight in the war for Europe. At the ferocious battle at Monte Cassino in 1944, Wojtek took an active part. It took four men to carry crates of twenty-five-pound artillery shells from trucks or ammunition dumps to the where the guns were emplaced, while Wojtek could carry one of these by himself – and apparently never dropped any. At the end of the war, he returned to the Borders, to Berwickshire, with the Free Polish soldiers who had decided to stay in Britain. Later he lived at Edinburgh Zoo until his death in 1963. There are many memories of Wojtek: plays, films, plaques, military banners and a splendid bronze sculpture in Princes Street Gardens.

24 July

Vast continents of clouds collided and the night sky crackled with lightning as thunder roared across the landscape. The storm began at 2 a.m. with bleary-eyed farce. The farmhouse is surrounded with motion-sensitive lights and these are often triggered by passing badgers, deer, sometimes even rabbits. When the first flash of sheet lightning flickered, it woke me, but thinking the sensors had clicked on I turned over. But then the light was followed by a low rumble that sounded very like a car engine revving and so I dashed downstairs and outside to apprehend the thieves, in my Y fronts, boots and an old anorak. Probably very scary. When lightning flared again, the farmhouse was suddenly framed in a flash photograph, and I shook my head, hoping no one had seen this crazed apparition, then made my way back to bed. On my way upstairs, I smelled burning and wondered about a forked lightning strike. But, in fact, I had left on a ring of the gas hob after cooking supper and it was slowly melting a plastic pan handle. Farce.

Ten or twelve miles above the farm, a very different drama was about to unfold. Mariners call the huge cumulonimbus clouds that cause thunderstorms by a different name. Perhaps because, above the ocean and its flat horizons, they can make out the developing shapes and colours of these masses of water vapour, they call them thunderheads. Yesterday's humidity and the enormous volumes of rising hot air had created a front, a long series of these clouds, what is known as a rainband. And at 2 a.m. the first of many detonations began. Most thunderstorms are brief, and, as the high stratospheric winds bring them in from a distance, children count between the flash of lightning and the faint rumble that follows to measure what that distance is. Then the dark eye of the storm appears overhead, is loudest, before being carried away by the wind. But for two hours the rainband brought many thunderheads and the angry sky flashed and roared, seeming to shake the house for a long time.

Because these huge clouds fly so high, their rain falls a very long way and is therefore very heavy, drumming on the slates before easing, then hammering down again. I worried about Wendy and her beautiful little foal, Echo, born in May. They are out continually now, in the paddock next to the house, and this was the first thunderstorm the wee one had seen. In a very grey light, at 5.30 a.m., I went out to check on them and both had their heads down, grazing.

Before the storm burst upon us, we welcomed another little one. At seven weeks, Tabitha is a tiny kitten, maybe only five or six inches long, and so that she will be safe, particularly from crows or buzzards, Lindsay has put her in an empty stable. Surrounded by small cardboard boxes, a milk bowl, a food bowl, some toys and a small blanket Lindsay had given to the breeder some weeks ago (so that it could acquire the mother's and siblings' scent), she seemed consumed by a nervous curiosity. Grace was entranced, constantly smiling as the little one skipped around. But this morning she was mewling, crying, not only because she missed her mother and her siblings but also because the storm had raged overhead while she was in a vast, unfamiliar place and all alone. Comfort arrived when Lindsay fed her and Grace has agreed to be her big sister. Her cat-name will be G-Cat.

25 July

At 6 a.m. the air was already heavy with humidity and temperatures here, at six hundred feet up in Scotland, are set to soar to eighty-plus Fahrenheit, perhaps ninety. But in the south of England it is forecast to climb to about a hundred degrees (close to thirty-eight degrees Celsius) – the hottest day ever recorded in Britain. More thunder will surely come tonight, and I think Lindsay will put Wendy and Echo back in their stable to avoid the heavy downpours and the threat of lightning strikes. Apparently the Met Office recorded more than 48,000 strikes in Britain two days ago.

This morning there was a dense mist lying over the Ettrick, but the climbing sun burned it off in the twenty minutes that Maidie and I were out. I noticed the first of the swallows' fledglings perched on the telegraph wire at Windy Gates. Eight of them were lined up as though it was their first day at flight school, waiting for the instructor to arrive.

26 July

A year after their encounter in the meadow, the anger of the Selkirk duellists flared into attempted rebellion. In February 1708, six French regiments and an Irish brigade mustered at Dunkirk, where they prepared to board five warships and twenty frigates. It was a credible and dangerous invasion force of about six thousand battle-hardened soldiers. And in the holds of the frigates they had stored arms for a further thirteen thousand men, the Scottish Jacobites, just like Walter Scott of Raeburn's family, who would flock to the royal standard when it was again raised. But there was a problem. James VIII and III had contracted measles. This delayed departure and may have softened morale. The plan had been to sail north to the Berwickshire coastline, then on to the Firth of Forth, and at a prearranged signal land the invasion force within striking distance of Edinburgh. With the focus on Britain's wars in Europe, James VIII and III and his advisors believed that there were only two and a half thousand regular soldiers, at most, based in Scotland and that the French regiments would easily overrun them. In the event bad weather, bad luck and bad judgement meant that the soldiers were never landed and the fleet had to circumnavigate Scotland before it could return to France.

Generally described as a fiasco, the abortive invasion of 1708 probably represented the Jacobites' best chance of success. In 1715 another rising fizzled out after the indecisive battle at Sheriffmuir near Perth and, while it met at first with spectacular success, the most famous Jacobite rebellion of 1745 was too little and too late – although at the time no one could know that.

After a shock victory at the Battle of Prestonpans on 21 September 1745, when the Highland charge swept the government army off the field in only fifteen minutes of fighting, Prince Charles set up his provisional government at Holyrood Palace in Edinburgh. Desperately short of money and supplies, and anxious to invade England before the winter, his commissioners sent out letters all over Scotland demanding support. Famous by this time as a town of shoemakers, or soutars, Selkirk was told to produce two thousand pairs of shoes. Up at Hartwoodmyres, the hill farm on our western horizon, William Ogilvie received a threatening letter. He was the Collector of the Cess, the land tax levied on landowners and farmers in the county of Selkirk, and if he did not immediately bring what receipts he had to Holyrood he would be 'deemed guilty of rebellion and military execution shall be complied against your person and effects. By His Highness' command'. To compound Ogilvie's problems, another letter was delivered to Hartwoodmyres, this time from the London government, also demanding the receipts of the Cess. As he looked out of his windows down our little valley, the Collector must have scratched his head and wondered which way the wind would blow.

No doubt after much thought, Ogilvie did nothing. He replied to neither letter, hoping his hill farm was too far from Edinburgh, and believing that he was certainly safe from a visit from a London official, at least for the time being. But on 14 October another letter from Holyrood arrived at Hartwoodmyres, not from an official but from a soldier, a cavalry commander. Robert Graeme of Garvock held estates in Strathallan in Perthshire and had raised a company of soldiers from his tenantry. On 15 October 1745, he led his men down the Long Track, riding at their head. There were perhaps eighty to a hundred Gaelic-speaking clansmen, armed with claymores, Lochaber axes, muskets and dirks, enough to intimidate any who saw them pass by the grass parks. If Graeme had reined his horse at Windy Gates, he could have seen Hartwoodmyres on its hillside and the line of the old Roman road that ran close to it.

After the clansmen had surrounded the house, Robert Graeme told William Ogilvie that unless he came to Peebles in two days' time with all the cash he could collect his goods would be seized, he would be summarily executed for treason and the clansmen would return to burn his house.

On 18 October, the Laird of Hartwoodmyres duly rode over the hills to Peebles and handed over £357, 15 shillings and 7 pence. Graeme issued a further demand, but by that time the year was wearing on and Prince Charles needed to lead his army on an invasion of England before the weather closed down all military options and his army of clansmen dispersed.

Split into three divisions, the Jacobite army finally advanced south at the beginning of November. Commanded by Lord Balmerino, the central division marched from Selkirk down the Long Track and on to Hawick and the Cheviots beyond. Led by their pipers, almost two thousand men wrapped in their plaids against the late autumn weather – Drummonds, Murrays and MacGregors from Perthshire, a squadron of Strathallan Horse, some of Balmerino's tenants from north Fife, and perhaps forty of Prince Charles' mounted Lifeguards, splendid in their blue uniforms – passed Windy Gates, moved down the track, crossed the Roman road and made their way south into England, praying that destiny marched with them.

A month later, after the retreat from Derby, groups of Highlanders were seen moving through the Borders countryside, some no doubt retracing their steps, crossing our farm, going home to the glens. As news slowly filtered north, it seemed to most that the rebellion had failed, but only just. The Highland army had advanced to within 120 miles of London, caused panic, and were only diverted by trickery and poor intelligence. After his capture at Culloden in April 1746, the only defeat the Jacobites suffered, Lord Balmerino was executed on the scaffold at Tower Hill in London, the executioner needing three blows to behead him. Robert Graeme escaped to Sweden. Up at Hartwoodmyres, William Ogilvie had to suffer much criticism for handing over the hard-won land tax to the Jacobite rebels, some of it from people who had fled the Borders at the time. But at least he kept his head and his house.

27 July

Cool morning rain rinsed the sticky humidity out of the air. All over Britain meteorological records have been tumbling. In the south of England the hottest day ever was recorded when temperatures climbed to 101 Fahrenheit in Cambridge. Edinburgh sweltered in 89 degrees, another record, while in the north-west of Scotland the night-time temperature was an uncomfortable 60 degrees. Here we had 86 Fahrenheit and a sweaty, sleepless 57 degrees during the night.

But on this silent morning the air was fresh at last, the scent of the honeysuckle near the Wood Barn intense and at the top of the Bottom Track tiny, vivid, red raspberries have formed on the canes in the last few days, as if they had been growing in a greenhouse. Which, of course, they have. I tasted one and it was soft but not yet sweet. The birds will spot them soon.

Streamers of mist clung to Newark Hill out to the west and the rounded summit of Huntly Hill looked as though summer snow had fallen. But then, moments later, the mist changed shape and, from clinging to the contours, it rose up and plumed the top like ice cream on a cone. Even with leaden rain clouds moving through and little light in the east, the land does not need the sun to look beautiful.

28 July

In 1708, the year when the first Jacobite rebellion failed, John Pringle returned to the Haining to contest the first general election to the parliament of the United Kingdom of Great Britain and Ireland. He need not have bothered to make the long journey north from London since no other candidate stood against him. His lucrative appointment as Keeper of the Signet allowed Pringle to develop the big house at the Haining. It grew into a U-shaped three-storey building with a large attic and an architraved doorway with the Pringle arms set above it. Wide policies were set out around the loch. This was a palpable break with the past. Good land had always been grazed or cultivated, but, enriched by their cut of the Equivalent (although there were rows about this – the Scots had expected to be paid in coin and not in Treasury bills, paper money still a novelty outside

London) and well-paid sinecures, landowners could afford to leave ground fallow around their houses, plant it with trees and shape vistas to please the eye.

Several big houses were built in the Borders in the early eighteenth century and for one of them excellent records survive. William Adam designed Mellerstain House near Kelso for the Baillie family in 1724 and, as at the Haining, he had a loch created in the foreground of a magnificent view. To finish it, he did what many English landowners did and had a folly built on a nearby hill. But it was the design of the house that showed how social attitudes were shifting, along with the redrawing of the political map.

Before the Union of 1707 and the creation of links with London, all Scots, rich and poor, titled and humble, spoke Scots and because of the influence of the Scottish Reformation and its emphasis on literacy and education, there was not a great gulf in social attitudes. But in the early eighteenth century they began to widen. In 1726 William Adam wrote to George Baillie, proposing to lower the level of the floor of the kitchen in the servants' wing of the grand new house to five feet below the windowsills. This would be 'so much better in that it prevents those in the kitchen and scullery from looking into the gardens'. Probably much influenced by her visits to London, Grizel Baillie set down thirty-seven different directions for her butler ('You must keep yourself very clean'), as well as detailing a system of signs to let him know when to clear away one course of a meal and bring in another and so on. Scotland was beginning to break up into more sharply defined social classes. And gradually, wealthy, titled people would increasingly be educated in England (Eton and Oxford and other combinations) and, crucially, would no longer speak Scots to their servants or tenants but a version of received pronunciation. The gentry were growing apart, becoming even more not like us.

With the clear-felling of the woodland at the south end of the loch, the Haining house has once again joined its policies, our farm. Something of the vision that the Pringle family had has been restored. This morning with Maidie I walked around the revealed landscape and its reopened vistas. And then suddenly – from nowhere, it seemed – I saw a woman in a white dress standing motionless by the lochside, gazing intently at the big house. Silence, stillness, a humid air, a freeze-frame. Suddenly a dog barked, Maidie answered, the woman turned, pulled up the collar of her long white coat and took a lead out of her pocket as a bay Labrador bounded along the path. For a few seconds I had seen a three-hundred-year-old tableau, a set-piece, not a Canaletto but more like Nicolas Poussin's *Et in Arcadia ego*.

The foresters were forced to knock down part of our boundary fence and now dog walkers are often seen in the Deer Park. We have horses there and plan to run cattle and sheep soon. We will need to fence it off securely for everyone's safety. It will be very expensive, but unless we do it the ground will be overrun and become useless for grazing. Then it will be *Et in Arcadia fui*.

30 July

Trailing wide, symmetrical wakes across the morning water, four ducks paddled up the Haining Loch as though they were part of a perspective drawing, leading the eye to the south façade of the big house and the vanishing point beyond. From out of the shadows, a heron flew low over the stillness, its slow wingbeats seeming insufficient to keep the great bird in the air. As it banked before the shore to find a perch, an extending claw touched the surface and a ripple sent out its concentric circles.

31 July

By 1753, John Pringle had begun to obliterate history and re-arrange geography. Having resigned his seat at Westminster to accept an appointment as a judge at the Court of Session in Edinburgh in 1729, he spent much more time at the Haining. No doubt well aware of the magnificence of the Baillie's new house at Mellerstain and others who had raised grand mansions in the Borders, he was determined to alter the map. Ordinary people, travellers, riders, carters and others, were not what he wanted to see from his drawing-room windows and so the Long Track was summarily erased. Perhaps deploying his legal expertise and political experience, Pringle had the road from Selkirk to Hawick shifted to the east, to its present course. Instead of an easy start to a journey through the flat Doocot Field for carters with a full load, they were forced to climb the Loan, a very steep hill beyond the town's South Port. And from there, there is another incline to the east of the Deer Park. The West Road from the Ettrick and Yarrow Valleys also came far too close to the Haining for Pringle's liking and it too was diverted by a more awkward route.

Selkirk was also too close, and the laird of the Haining bought up houses in the Peelgait, what is now Castle Street, and demolished them to create a cordon sanitaire between his family and the common herd.

But, behind all of this topographical bullying, problems were lurking. Despite his preferment at Westminster and almost twenty-five years on the bench of the Court of Session, Pringle was in financial trouble. Debts had piled up so high that when Pringle died, his son Andrew refused to accept the estate as his inheritance. Instead, he sold it to his younger brother, also John Pringle. Like Mark the duellist, he had made a fortune in trade, as a merchant importing wine from Madeira.

The sum of John Pringle the merchant's ambition can be seen on a fascinating map, the first map of our farm and its surroundings to survive. *A Plan of Haining Estate. John Pringle Esq, Proprietor, accurately suirvy'd [sic] & drawn by John Scot* is in fact nothing of the kind. Both the Long Track and the Deer Park and its pale have been summarily and completely effaced, even though traces of them were certainly extant. Fanciful and banal field-names have been scattered over the landscape. Moss Sluice Park, Cow Park, Beach (presumably a scribal error for Beech, the sea being at some distance), Hill Park and most of the others have all long since disappeared for the excellent reason that, apart from Pringle and his surveyor, no one called them that. Our land is labelled as East Haining, the Pond Park, Ryegrass Park (a hint that the new strains of grass had been sown) and the Back Park. Parks also imply grazing land rather than any cultivation. The sole echo from the long past is a misappropriation. The course of the Hartwoodburn is plotted more or less accurately, but called The Lake, a memory of the Lost Loch where hunter-gatherers fished, fowled and knapped their flints. Other names are entirely unhelpful, but probably more accurate, such as The Wood, the Meadow Spot or The Whins.

Most striking is a long perimeter of what seem to be strips of woodland around the policies closest to the big house. Like a huge screen, it runs around the eastern boundary of the Deer Park, up the line of the Long Track, west of the Doocot Field, and then returns to the Haining slightly to the north. It resembles a vast suburban hedge designed to keep out prying eyes or unwelcome visitors. Pringle must surely have known that 'haining' is an old Scots word for a hedged enclosure.

It seems to me that John Scot had drawn a fantasy landscape, what seems more like plans and aspirations. A later map, drawn almost exactly a hundred years after the first one, has reinstated all of the features that had been obliterated on paper. The Long Track has acquired an avenue of trees, the Deer Park has its old quarries accurately plotted and there is no sign of the vast hedge. It may not ever have been planted.

By 1757 the medieval map was fading fast, as fields were enclosed and open grazing was shrinking as lairds used the law to take more and more acreage into private ownership. There is no note of even the modern limits of the Selkirk Common on

Scot's map. The Haining estate extended much further to the west, beyond the motte (spelled *Moat* and possibly represented as a version of a folly) to Howden Farm and the mouth of the Ettrick Valley.

The map is intriguing not only for what it omits but also for what is included. In addition to *Mongos Well* (sic), St Mungo's Well in the Deer Park, there are others plotted that seem to have disappeared. Most suggestive is a well precisely where Walter, Rory and I found the remains of an early medieval village in the Doocot Field. And by the Hartwoodburn there is a Castle Park and a Castle Ford. They seem not to relate to the motte. Perhaps the names are an invention, an attempt at creating a setting of fake antiquity in the shadow of a place that was genuinely ancient.

Most surprising of all is another disappearance. On the 1757 map there is no trace of our farmhouse. When we had it rebuilt and extended, the masons took off the old roof and found under the slates adzed tree boughs rather than machine-cut rafters. And the pitch was very steep, usually a sign of a thatched roof and the need for faster run-off in wet weather. The house seemed to be older than the 1821 date we found on the sandstone plaque, but on this map there was no sign of it. Perhaps this was more evidence of a localised clearance, as John Pringle pushed ordinary people to the margins of the view from Haining House.

August

1 August

While wealth and privilege were changing the look of the land for pleasure – or at least to keep up with the neighbours – farmers, farm workers and a brilliant blacksmith from Berwickshire were changing it so that it became more productive. In 1764, workmen built a great forge and a range of workshops at Blackadder Mount Farm near Duns. Funded by John Renton, a visionary landowner with more than a good view in mind, a young blacksmith began work on an invention that would change the world much for the better. The Old Scots Plough was very inefficient. Made from wood and dragged through the unyielding ground by a team of four oxen or heavy horses, it often broke down. It needed not only a ploughman to guide it and a goadman to urge on the beasts, there was also a small army of plough followers. They had to beat down big clods with a mel and pull out the weeds. In his new smiddy, James Small was determined to create a new plough, one that was immeasurably more efficient. And he succeeded.

Giving the ploughshare a new, screwed shape, much like modern ploughs, and casting it all in iron, Small's design reduced friction, delved deeper and turned over the furrow slice so completely that weeds were buried and became mulched as extra fertiliser. There was no longer any need for plough followers or goadmen, and one skilled man could do everything. Guiding two strong horses (and later one, when the breeding of Clydesdales began in earnest as a response to Small's invention), he could wrap the long reins around his fists while holding the plough stilts straight. Farming was revolutionised by James Small's genius. Because deeper ploughing meant better drainage and fewer workers were needed, much more land was brought into cultivation, and crop yields soared.

The grass parks to the west and around the farm of Hartwoodburn shown on the 1757 map were much larger when the 1858 map was surveyed. This was to allow longer furrow-runs with the new plough and reduce the wastage of land around the margins of smaller fields. While our farm is still shown as all grazing in the later map, with liberal tree planting, the plotting of several sheep folds suggests that flocks were reared close to Haining House. Instead of being used to feed a sense of acquired grandeur, the land was being used to feed people.

2 August

The first hour of each morning is more or less exactly the same throughout the year, the only difference being the seasons, dark in winter, light in summer. Once downstairs I pick up the three feed bowls for the dogs, change their water bowl, give Lily lots of pets and attention (she is old and sensible enough not to need to sleep in a crate) and go into the kitchen to measure out the breakfasts. After making tea, I take out each one for a morning pee. Once that is done, the dogs eat their breakfasts and I do a programme of stretching and bending, more and more necessary as the years wear on. Then I water the tomatoes and lettuce in the conservatory. I open my laptop to check the weather, and also check we are not broke, or no more broke than usual. Then I have my breakfast before taking Maidie out for our early morning walk. I do all of this in the same order, like a series of reflexes rather than any conscious thought. Probably just as well.

3 August

On either side of the Long Track run parallel strips of temperate
jungle, tangles of cow parsley, different grasses, mostly tall
Timothy, nettles, thistles, wild parsnip, poisonous hemlock,
willowherb and many leaves I can't identify. On this misty
morning, the dried and darkening crowns of the cow parsley
and parsnip were enmeshed by hundreds of spiders' webs. Silver,
dew-drenched, their fine lace tracery mummified the dying
plants. White thistledown lay thick in places, marooned on this
windless morning.

This is the seed time, when parent plants cast their fertility.
With no breeze to carry away next year's growth, the grasses
and the parsley need the goldfinches and the other tiny birds
to work for them. When Maidie and I came to Windy Gates,
dozens of little ones were pecking amongst the stones of the
track, searching for seeds. Beyond the jungle strips, fat lambs,
belly-deep in lush grass, watched us pass, and unseen, down on
the lane, I heard the clip-clop of a trotting horse.

4 August

Last summer's storms blew down a magnificent beech tree in
the Deer Park, and yesterday two foresters logged most of it.
They had to leave about twelve feet of the main trunk because
if they had cut any more it would have altered the balance and
the tenacious, sinuous roots on one edge might have sprung it
back up to the vertical, not something you want to have happen
with a chainsaw in your hand. With a tractor and tipping trailer,
the foresters brought down about twelve tons of prime hardwood and dumped it at the Wood Barn. Most of it is seasoned.
The big discs from the upper trunk and main limbs will take
some splitting but there are plenty of smaller logs and all cut
to a handy size.

Many, many butterflies flittered around Maidie and me this morning and the flies were at their worst. The swallows would say they were at their best and tastiest as they swooped around us, scooping up an aerobatic breakfast.

6 August

Indefatigable, precise and persistent, Rory keeps digging up more and more of the history of here. At the weekend he found a Charles I half-groat in the Doocot Field, probably minted in 1645, the same year as the slaughter at Philiphaugh. Was it dropped by a soldier? Maybe. The tiny coin made the metal detector buzz only three feet from the doorway of the doocot itself and, in the addictive sequence that followed, one I recognise all too readily, that find set Rory off researching the history of what the early maps of the Haining call a pigeon house. And when he sent me what he had found, that set Maidie and I off on a walk over to the doocot.

As Rory pointed out, the building was designed in the lectern style, very singular and peculiar to Scotland. The back wall was sheer with no openings and faced north, backing the doocot into the bitter winter winds. On the southern façade a steeply pitched roof sloped down to a lower wall where the doorway once was. And on either side there were smooth walls topped by crow-stepped gables. Ingeniously worked out, the lectern shape was designed to keep out predators. When Maidie and I reached the old ruin, I could see the remains of a projecting string course that would stop rats, stoats or pine martens from climbing the walls. The roof was so steeply pitched and probably smoothly slated that, even if a rat was acrobatic enough to climb around the string course and get up the wall, it would slide off the roof, its claws finding no grip. Pigeons could perch there and sun themselves, but the pitch was too steep for heavier hawks or buzzards. The entrances or flight holes were usually few, only six or so, and they were almost always placed in the centre of the pitch. Doocots were often built in open spaces, just like the field we stood in, and not near trees where hawks could conceal themselves.

Inside this fascinating little building were many nesting boxes built into the walls, about a foot square in shape and formed with flagstones. On the upstanding back wall I counted two hundred, and on the other three walls there will have been more, perhaps five hundred in all. It occurred to me this morning that there was less of the doocot than there used to be and later, in my office, I checked some old photographs. Vandals or thieves have clearly paid several visits because until recently part of the east wall and its nesting boxes, and parts of the south wall, were still standing. But around the base of the wall I could see no fallen stones. Had people been robbing out the sandstone? Garden rockeries?

As Rory noted, doocots were popular on farms and estates because they were a year-round source of fresh meat, eggs, a rich guano that could be used to make lime mortar, tanning fluid or be an ingredient in gunpowder, as well as feathers for mattresses or pillows. Whoever was sent over from Haining House to harvest pigeons had, literally, a shitty job. Since the birds will not fly if they cannot see, death came to the doocot in the dark. Climbing a ladder, the unlucky servant would either lift squabs (unfledged chicks) or adult birds off their nests and then wring their necks. And just as human beings do when the trapdoor on the scaffold opens, the pigeons will have defecated when they were killed in this way. Most doocots were not populated by the ubiquitous wood pigeons that pester us around the farm. Their meat was dark and thought to be too gamey. Instead, white rock doves were preferred. More ornamental, their meat was also white, like chicken.

The doves fed themselves but could eat so much ripe corn that farmers complained. In fact, it was said that the doves from aristocratic doocots, *colombiers*, eating so much at harvest time was one of the sparks that ignited the French Revolution. In 1617, the number and size of doocots in Scotland was restricted by law and any more that were built were to be no more than two miles from the owner's house – so that they ate his corn.

Visible from the windows and terrace of Haining House, and probably commissioned by John Pringle after 1729, the doocot had a secondary function as a kind of folly, a building that filled a western vista. It also became a monument to change. When butcher meat became cheaper and more readily available as farming modernised, and new industrial processes dispensed with the need for feathers and guano, doocots fell into general disuse. After the middle of the nineteenth century few new ones seem to have been built, and most of those for aesthetic rather than practical reasons. Which must have come as a relief to some. Harvesting the birds and eggs was one less shitty job that needed to be done.

7 August

The Fat Lambs were all lying down this morning, with grass up to their chins. The blackfaces are a strange black blank because their eyes are also black and so their features look like a mask, only the big, flicking ears showing any animation. Some have grown so fat on the lush grass you could roll them like a barrel.

Farming time ticks at a different pace from timetables, nine-to-five days and weeks, TV schedules or any of the everyday metrics that govern our lives. Growth proceeds at its own pace, influenced principally by the weather. The lambs are fatter earlier because the damp, warm and frequently humid summer has made the grass grow thick, sweet and tall. Farming time is accelerating but soon, as the days begin to shorten and the nights turn colder, it will slow down. And then, in midwinter, it will stop, dead.

Meanwhile time's arrows flew overhead. This morning we seemed to be under the flight path into Edinburgh Airport and two aeroplanes were beginning their descent, their silver underbellies lit by the sun. The loud drone of the engines made Maidie stop sniffing and look up, the tiny dog searching the sky for the distant roar of the leviathans. One of the planes banked slightly and I could make out the Union Jack livery of British Airways on its tail.

9 August

Heavy, continuous rain began to fall just before 6 a.m. and looks set to last most of the day. Scotland appears to have slipped down through the latitudes into the tropics: sun, humidity, thunder and monsoons. What these conditions will do to the imminent harvest, I don't like to think. Further down the Tweed Valley, some farmers have cut their corn but here, six hundred feet up, we wait for better weather.

10 August

The wild raspberries at the top of the Bottom Track are plump with summer rain but still not sweet. Many have survived the birds' attentions, probably because the hips and haws of the rowans and the thorns are turning red, and the geans, the wild Scottish cherries along the Top Track, look like little green apples. There will be an autumn feast for all.

Today's forecast is for more heavy rain and we find ourselves having to cope with persistent bad weather, something we expect in the winter but not in the summer. Wendy and her foal should be out all the time so that wee Echo can stretch her long legs and thrive on the sweet grass, but both have to come in because of the thunder and the danger of a lightning strike.

11 August

In 1792, John Pringle the merchant bequeathed the Haining estate to his great-nephew, Mark Pringle. The grandson of the duellist, he had ambitions for his inheritance, especially the house. As the trees planted by both John Pringles matured, the policies grew more pleasing with every passing year. But the house at their centre, what the loch and the trees around it were intended to frame, did not fit with Mark Pringle's vision for his new estate. In 1794 work on a new house began. Georgian in style and scale, with neo-classical columns and a portico to the south, its look would sit better in the newly created Romantic landscape than the old Scottish vernacular pile. Postcard-perfect, picturesque Poussin was preferred to the snug utility of the past.

Views of the house, as well as those from it, were carefully considered and Windy Gates was thought to offer the best aspect. The Long Track was to be lined with specimen hardwoods and known as the South Drive. It was to become part of a circuit for genteel afternoon outings. When gentlemen and ladies with their parasols climbed into a well-sprung carriage on a sunlit afternoon at the big house, it took them first around the Haining Moss to the east. The hard standing for the drive is still there, running along the fence line. Then the coachman turned the carriage south down a lovely track that still runs through my neighbour's grass park to the west of our farmhouse. In places it must have been a business to make because the road bed is shelved into the natural undulations of the field. Once they reached the lane, the line of the old Roman road, the reins were flicked and the horses guided to go left until they reached the foot of the Long Track/South Drive. Once up at Windy Gates, the best, most idyllic view of the new neo-classical house could be enjoyed. *Et in Arcadia sumus.*

With the new felling to the south of the loch, that old vista has opened once more and now, since the Haining and its grounds have been bequeathed to the public for their use, everyone can enjoy it. And now one only needs to move past the gas substation with its spiky railings, safety warnings, shiny pipes and other mysterious machinery to return to something first seen in Regency times, the epitome of fashionable elegance. In his otherwise encyclopaedic and scholarly *History of Selkirkshire*, Thomas Craig-Brown is gushy and pretentious about this vista:

> In early summer, when the sun is shining, and when the loch reflects a sky of blue, the wanderer by its margin might well believe himself on the enchanting shore of Como, or roaming by the lake of the Doria Pamfili Palace at Rome.

But there is one important, and remarkable, qualification to this puffery. The Pringles did not demolish the despised old house and kept it standing close to the new so that it could be used as servants' quarters. Not quite such an idyll, and not something that would ever have been considered at the Doria Pamfilj.

12 August

Piou-piou is as close as I can get to the plaintive cry of a buzzard, heard as Maidie and I walked through the rain down the Long Track. It has been falling for more than twenty-four hours without cease, often heavily, and the land was sodden, the long grass bent, the leaves dripping and the track a brown torrent. Perched on the telegraph line was the buzzard, the hawk in the rain. On this soaking morning, it searched the margins of the track for movement. But there was none.

The Haining Loch looked less like *Il lago di Como* and more like Loch Lomond in January.

15 August

Even though I have ridden only occasionally, and more than ten years ago, I believe I have a folk memory, an innate feeling, perhaps even an instinctive liking, for horses. As late as 1800 most people worked on the land or in food production in some way and they were constantly close to horses. They pulled ploughs, carts and were ridden. We have been in the company of horses for thousands of years in one way or another and it is only in my lifetime that they have been replaced by the internal combustion engine.

In my youth, Kyle's Stables were right in the centre of Kelso, a set of rickety green looseboxes just off the Square. In what is beyond historically ironic, a car park now occupies the site. Around the town, horses pulled carts loaded with bags of coal, milk floats and much else. They were ridden, and still are, for pleasure in the Border common ridings and Kelso has a famous racecourse. This departure of the horse from our lives is a profound break with a long past and it fills me with pleasure that we have them here. Such beautiful, noble, trusting creatures, they deserve our gratitude, love and respect. They are not quite yet fled from our lives.

With some trouble (lots of medical and police checks, and some bureaucracy, and a cost of £70), I renewed my shotgun licence, and hidden away in a locked cabinet are two excellent old-fashioned side-by-side guns. Beautifully made with tooled metal and shiny walnut stocks, they might become important to the farm and my efforts at growing our own food. We may have to shoot it as well. There are far too many wood pigeons and pheasants, but the rabbits who scamper about in profusion might be safe enough. For some years, I have seen diseased rabbits dying in the verges, probably infected by new versions of myxomatosis. The other day I saw one sitting on the white lines in the middle of the main road, clearly sick, and I hoped a big lorry would put it out of its misery soon. I couldn't find the courage. It would vex me very much to shoot birds, to say nothing of gutting them and plucking them, but who knows, it might come to that. Adam is an excellent shot, a natural, far better than me when we used to go clay pigeon shooting. I should buy some new cartridges, just in case.

The first leaves fell today. But the land is not gently mellowing into autumn. Instead the weeks of rain have left the ripening fields sodden. The combine harvester churned so much mud in the barley fields that it was forced to stop, leaving the steeper swales and hollows for another, drier day. If it had attempted even these slopes, the great machine might have slid and crushed the standing crop.

17 August

At the turn to the Top Track, the early morning sun lights the grass like an invitation. At last there is a little warmth as Maidie and I splash through the puddles up to Windy Gates. The accidental apple tree is laden and I noticed some finches pecking at its ripening fruit. When the birds get to the core and the shiny black seeds, they will carry off their cargo and drop it far from the mother tree to make new life possible.

18 August

When we left the house, the sun was shining brightly, but by the time Maidie and I reached Windy Gates, the rain was sheeting down. August. Without a hat, I pulled my jacket tight and marched on briskly. Shelter from bad weather is something fundamental and many of my instincts as a parent and a husband are linked to the need to provide protection. Sometimes in dreams, all five of us are caught in a storm and we find somewhere to get out of the worst of it. The image is always the same. With my back to the weather, I keep the rain or snow or wind off Lindsay and the children. It is not heroic, just instinctive. The reality is that at our advanced years it should be the children protecting us. But instinct has no respect for reality.

Thinking about this, I was reminded of a story about how parenthood never ends. A ninety-five-year-old lady was being interviewed and asked about what made her happy or anxious. 'Well' she said, 'I can relax a bit now that I have got my eldest into a care home.'

19 August

After only a handful of weeks when early morning walks did not need a jacket, I wrapped up warm even though the sun shone. A stiffening breeze was blowing out of the east, drying the drenched fields, bringing autumn closer. The tracks are still speckled with the tiny browning petals of the blossom that brought the hips and haws that are bending the branches of the rowans and thorns. Soon their leaves will follow, as the year and the land begin to die.

In yesterday's newspaper I saw a very beautiful photograph of the wild landscape of the Isle of Lewis. It looked south to the hills of Harris. What the journalist (also the photographer) described as 'layer upon soft layer of pastel colours, from gold through green to blue, laid like chiffon scarves across the horizon'. Wonderful writing that caught a moment of quiet subtlety in that elemental place. I began to think of Dalmore, somewhere my thoughts sometimes turn. I remembered the funeral at the graveyard, the smooth, wave-worn boulders on the beach and the echoes of a different cycle of time I first heard whispered there. The article in the paper reminded me how awed I was by the mighty majesty of Dalmore, how the thunder of the waves and the eternities of the Atlantic stirred a sense of the vastness of creation.

But I do not love it. The beauty of the bay, the breakers, the sea stacks and the cliffs are cold and perfect, untouched, un-affected by the hand of men and women. For me, it is the fields that are beautiful, the hedges, the woods, the shelter-belts, the tracks, the fences, and the corners where I might sit to watch the birds, our horses and the glowing sum of all the work done over millennia by hundreds of generations of farmers who shaped and tamed the wilderness, the prehistoric Wildwood, into places where food could be grown. The warmth of domes-ticity, its detail and connectedness are what I love and where I feel my heart is at home.

21 August

For reasons I must discover from my neighbour, he has put two yellow ewes in one of the grass parks by the Long Track. Bright primrose yellow, with red numbers on their rumps, they look very odd indeed, like cartoon sheep. So odd that the normally placid pony Princess spooked when she was led past them and Saffy, an old cob who has seen it all – but never seen anything like these yellow ewes – went sideways as she passed.

In Huppanova, the old bull is back with his cows. A lugubrious giant, his balls swinging under him like the pendulum of a grandfather clock, he seemed to be building up his stamina, munching great mouthfuls of lush grass. His chest is so deep and his great head so massive, it will be a feat of extraordinary athleticism when he starts to mount his cows. I just hope that not too many of them decide to move forward at the last moment. Some are still suckling their calves from this spring and they might not be too keen on the old boy's attentions.

Given the difficulty the older cows can have when they birth, I would not be surprised at diffidence. A few years ago, my old neighbour called me to ask for help pulling out a calf that was breeched, positioned the wrong way round. As a last resort the vet was planning a bovine version of a caesarean section, 'going in the side door', but they wanted to try one last heave. Lying slightly on her side, the old cow was distressed, lowing loud and plaintively. While my neighbour did his best to stop her kicking me, I plunged my arm past all sorts and into the cow's uterus. I was given the job because I have long arms, longer than the wee vet's. After feeling around at his instruction, and being more or less continuously shat on, I felt I had hold of the calf's lower jaw, its soft, rubbery teeth, and I pulled as hard as I could. I must have been able to turn her, for she shot out, shoving me backwards into the straw and landing on my legs.

After I'd had a wash by the tap in the steading and a tin of cold lager, I saw the old cow get up and start licking her girl. My other reward was to choose a name. Daisy, of course.

22 August

Lindsay has decided that the Deer Park needs to be properly fenced and work at last started in earnest yesterday. Surrounded by a broken perimeter of ancient and rusty barbed-wire fencing, a run of medieval ditches and revetted banks, with a high but porous dyke on the eastern side, the park had become useless for grazing. And so, for what will be a small fortune, hundreds of fence posts will be knocked in, stock mesh and barbed wire attached and an electrified wire run through some of it. The voracious appetites of many cows are urgently needed to reduce the tangle to manageable pasture. What we own is beautiful, but it is not scenery and it needs to be grazed and maintained. And that costs a great deal of effort and money.

On the old western road to Selkirk, the one that was shifted north by the Pringles, Rory has found something exciting, possibly the closest thing to real treasure. It is a cast copper alloy fragment of what is probably an Anglo-Saxon brooch, perhaps sixth or seventh century. That is very interesting and another small part of a building, developing mosaic, a picture of early Selkirk that is slowly coming into focus.

23 August

A tiny, velvety, perfect young rabbit, no more than a few weeks old, sat stock-still in the long grass where the Bottom Track turns. I saw it late, when Maidie and I were almost upon it, only five yards away, and all that moved was its blinking eye, its heart no doubt surging with terror. And yet, the fierce Westie, always avid for every sniff of potential prey on our morning walks, did not notice and ambled on along the Top Track. The little rabbit stayed motionless until we were about twenty yards away – and then resumed grazing! Animals truly do live in the moment.

A breezy sun this morning and a good forecast for the next few days. No sign of the tropical downpours that have drenched the land for most of August. Perhaps the bank holiday will be warm and the roads of the south of England will be cheerfully clogged as city dwellers try to drive somewhere else.

Maidie is becoming occasionally hysterical (not literally, of course, since she was spayed a few weeks ago), barking, rushing at and nipping the other dogs, trying to get loose to charge at the foal in the home paddock. Sometimes for ten minutes or more she simply loses it. I grab her, shout commands, and she is cowed and calm for a few seconds before another firework goes off in her head.

24 August

The morning glories of our little valley glowed in the sunshine. Summer had come back. But instead of working outside, working up a sweat, I intended to spend the morning in complete black-darkness in the bowels of the earth. Walter had sent me a remarkable photograph of a tunnel under the courtyard in front of Haining House that led to a deep ravine, the Clock Sorrow, the stream that drains the waters of the loch into the River Ettrick. Believing discretion to be the better part, Walter (or more likely Evelyn, Mrs Elliot, advising that it definitely was NOT a good idea for an eighty-four year old) declined to join the expedition, and Rory Low and I agreed to meet and explore.

And immediately we went the wrong way, completely missing the entrance. Scrambling down into the Clock Sorrow, we found another opening in the steep bank, what we thought was the beginning of the tunnel. With well-made lintels and jambs of cut and dressed sandstone, and a broken set of steps that led up to the courtyard, we were sure this was it. Not believing in discretion, and not willing to wait for the younger and fitter man to lead the way, I crawled across the mud that had piled up at the entrance. After a yard or two we were able to stand upright, phone torches in hand. And a step further was very nearly a step too far. Rocking forward slightly with my momentum, I found myself standing on the damp, slippery and sheer edge of a deep pit of fathomless darkness. The torches showed it was perhaps twelve or fourteen feet deep, with a bevelled bottom full of rubbish and an eerie echo. Above it was a perfectly domed roof. The air was chill and clammy at the same time.

This was an ice house. When grand mansions like the Haining were built in the eighteenth and nineteenth centuries, many had these curious, windowless, tomb-like structures erected close at hand. Ice will have been harvested from the loch in blocks during the winter (or bought from Norwegian suppliers, if the winter was mild) and stored in this great pit with the bottom rounded like a basin to gather meltwater and packed with straw for insulation. The ice was not used to make drinks cold but to act as refrigeration to keep meat, fish and other perishables from spoiling. Like many of these new mansions, the Haining was built for entertaining and large quantities of food will have been needed from time to time.

Having failed to find the entrance to the tunnel and avoided falling into the ice house, we crawled back out and walked up to the courtyard. Preparations were being made for a wedding and a marquee had been erected next to the big house. Ian Bradshaw chairs the Haining Trust and he showed us where the entrance to the tunnel was. Rory scrambled up the side of the Clock Sorrow, pulled away the branches that had hidden it – and pronounced it too dangerous to enter. But disappointment was mercifully short-lived because Ian offered us a tour of the cellars of the Georgian house. It was fascinating.

More evidence of entertaining on a large scale could be seen in the scores of large wine bins that once held thousands of bottles. Since at least two Pringles had made money in Madeira, it was safe to assume a knowledge of what was good and where to buy it. Under the entrance to the mansion were some much older walls. Built mainly from whinstone (like our farmhouse) with lintels and jambs of dressed sandstone, they were very intriguing. Ian showed us where the entrance to the tunnel had been walled up and there was a small mousehole drain leading into it.

Walter's photograph from the *Scotsman* of 8 March 2005 shows a massive construction. No mere drain for water from the loch, this had been built to a height of more than six feet and was two feet, six inches wide. The floor was paved with rounded cobbles, probably excavated from the bed of the Ettrick, the walls were very thick and the roof vaulted. Running for one hundred and fifty feet from under the entrance of the main house, it terminated in the deep declivity of the Clock Sorrow. But what was it for?

The house on the map of 1757 turns out to have had an ancestor. Haining Tower was the style of fortification known as a peel tower, characteristic of the Borders in the late medieval period, the time of the Border Reivers. Built before 1463 and used until the successor house was raised after 1625, the tower was held by the Scotts, a branch of the reiving family that eventually became the Dukes of Buccleuch. These families had to survive in what was essentially a criminal society and the tower was designed to be defended. If attackers looked numerous and likely to be persistent, waiting to starve out or burn out the Scotts, then a means of escape will have been thought useful. Perhaps under cover of darkness the defenders might have made their way quietly along the tunnel, out of the narrow and probably well-hidden doorway in the bank of the Clock Sorrow and down its deep ravine to the Ettrick. Someone who was surefooted and knew the ground well could easily have fled through the tunnel and escaped undetected. It would also have been a useful place to conceal stolen goods and contraband.

When Ian took us back up to the outer courtyard, he told us that a digger working in the courtyard had accidentally skinned the roof off the tunnel and collapsed part of it. It lay only three feet below the surface. That led to a theory about its strange and elaborate construction. Rather than tunnel from the bank of the Clock Sorrow, it is likely that the Scotts dug a deep trench, built the tunnel in it and the covered it over. A dark place built in dark times.

25 August

The faint purr of a quad bike up on the ridge above Brownmoor was a cheering sound, a joy to hear someone up and doing on a still, sunlit morning, checking the ewes and lambs. And on that clear morning the vistas from the ridge must have been long, perhaps already hazy. Diamond-glinting with dew, a trail of thistle-down had been caught by the grass at the side of the track at Windy Gates. Like the spoor of a giant snail, it wound around the old road and out of sight. There must have been a night wind. As we passed on our way to the Deer Park, a posse of fat calves, their condition glistening in the warming sunshine, trotted over to the electric fence to stand and stare, gormless, at Maidie and me. Normally fierce, the little terrier hid behind my legs.

Yesterday afternoon Grace came into my office to show me a locket her mum has let her wear. It divides into four folding hearts, each one holding a tiny picture. Grace put the locket in my hand, pointed at the pictures and recited, 'Mummy, Daddy, me and Hannah'. For a moment I could not catch my breath. For five years of grieving, I have not had the courage to look at photographs of Hannah. But there she was. An image less than a centimetre square, its power overwhelmed me. A tiny head of very dark hair and swaddled in a baby blanket. That was all I could see, and all I could think of was the blanket. It should have kept her warm, but underneath its folds Hannah was cold.

For a day I have carried that image in my head, tears prickling often, even as I write this. I have lost the rage at fate, at ill fortune and all of the easy conclusions, but the pain remains and will always remain. I wonder what Grace makes of the tiny photograph. I did not let her see my tears.

26 August

Much of yesterday and today were spent in Edinburgh at the book festival. I gave a talk to about seven hundred people in the main marquee on my book about St Cuthbert and Lindisfarne. It was the first time I had spoken to an audience about it and I was not at all sure about how it went down. But when I walked around the festival site today for about twenty minutes five people approached me to say it was good, even 'brilliant'. That was a relief.

I had some warming moments of a different sort. An Irish journalist I have known for forty years could not come to my event because of failing eyesight but he sent me a letter and a card to thank me for help I had given him in 1977! Remarkable and moving. An old lady from the Isle of Lewis comes to see me each year and we converse in Gaelic. When I kissed her, she clapped my cheek and smiled up at me. From Melrose, a lovely lady who teaches at the primary school brought her great-aunt, Margaret. She is ninety and felt that this was the right time to adopt me.

27 August

The mystery of the yellow ewes is solved. They have been dyed that strange colour so that they can be shown at an agricultural show. With a certain air of disdain, or perhaps puzzlement, my neighbour told me that they were not his, definitely, certainly not his, and that he was grazing them for a friend. Apparently sheep are also dyed red and brown for shows. It seems daft to me.

28 August

I have grown fatter. Catching a wince-inducing sideways glance in a full-length mirror this morning, I noticed that without the camouflage of a baggy shirt my belly was sticking out. Out the front. Not a good look certainly, but also a relic of an ancient cycle. When all harvests were in at the end of the summer – corn, garden vegetables, wild berries, fruit, early morning mushrooms and much else – my farm labourer ancestors ate well, better than at any other time of the year. It was a diet my grannie Bina understood and passed on to me, unfortunately. Based on carbohydrates and fat rather than much meat, my people ate porridge with cream or salt and butter, bannocks, scones, dumplings, blood puddings, pies both savoury and sweet, cheese, oatcakes, butter and more baked goods.

Bina was also fluent in the language of soup. Always in the plural, she would say of the ham bone stock and vegetable broth she always had seething on the cooker, 'They're better on the third day' or 'Aye, they're good but need salting.' More like old-fashioned potage, her soup had many and any ingredients, so that it really could only be thought of in the plural.

This diet of the old life, the farming life Bina left behind, is an unwelcome inheritance for me. My people spent their lives in the fields, in all weathers, and they needed as many calories as possible to sustain them. I sit on my backside to work and if I need to lose weight I simply stop eating their diet. But when I splod out two tablespoons of low-fat yoghourt over my seasonal berries (seasonal in Spain), out of the corner of my eye I can see Bina shaking her head.

Bina was born at the farm of Cliftonhill, near Kelso, in 1890 and it often seems to me that my family has walked around the edge of a circle that leads from her cottage to here, one hundred and thirty years later. Just as she did, we dance to a different music of time: the grass clock, horse clocks, dog clocks and the all-governing weather clock. In contrast to the problems of my expanding waistline, the animals who live outdoors need to eat as much they can so that they store fat for the lean months of winter. When I check the horses in the outbye fields each morning, even the oldest of the Old Boys look sleek. The grass clock is running slow because of the perfect growing weather we have had, and it will last long into the late autumn. Calorie derives from the Latin *calor*, for heat, and the Old Boys are busy stoking their internal boilers as they munch their way methodically through the juicy, sweet, rain-fed grass.

I find these ancient rhythms consoling, an antidote to the countdown of the months and years.

I watched a tiny acrobat performing this morning. A grey squirrel ran along the top rail of the gate into the Deer Park before skipping along the fence, turning with a flick of its bushy tail and running back along the gate. It seemed to me to be sheer exuberance.

29 August

As summer draws in, the swallows are preparing to leave, feeding on the abundant flies to store energy, the young chicks bulking up for the long and perilous haul to the South African sun. Soon they will begin to perch in long rows on the telegraph wires, massing, perhaps even bonding, before their flight south. Our parent swallows certainly recognise their own chicks, and born in the same small stable yard it seems likely that the different families of swallows know each other. Perhaps this is fanciful, wishful, but I like to think of them on their epic journey, flying two hundred miles a day, as the Henhouse Gang. How they find us again each spring is a magical mystery to me.

30 August

Worlds away from it and yet intimately connected to the life of our little farm, I found myself in Edinburgh touring book-shops to store up winter reading. Pure pleasure, and a heavy bag.

31 August

Sometimes I think I allow dog poop to influence my mood too readily. If Maidie does one during our morning walk, and again when I take her out after her supper, that is a good day. If she is distracted and refuses or forgets to perform and I have to clean up the consequences off the floor, scrubbing and spraying, that is not a good day. As I stare at my dog's back end, my world feels as though it is shrinking.

September

1 September

Grace has recently taken to painting small stones. Bright ultramarine blue, hot pink, white and red seem to be favourite colours. When she completes a batch, Grace then places them carefully at the foot of an upstanding feature: below the stone birdbath on one of the lawns, at the side of the Wood Barn, behind a water bucket, or by the purple acer at the gate into the stable yard. She must wait until the coast is clear, when Lindsay and I are in the house, before the ritual of placing takes place because I have never seen her doing it. What archaeologists will make of this in two hundred years' time I do not know. Perhaps only Grace does.

The first day of autumn sees the magical emergence of funghi of all sorts. The overnight temperature must have been optimal. On the margins of the Top Track and on the lawn by the birdbath, I have seen four sorts. A tall white variety known as lawyer's wig looks particularly evil. It is classified as Coprinus, which means 'living off dung', and while it is edible, it should not be consumed with alcohol. If you eat lawyer's wig with a glass of wine, you will vomit. Perhaps one day we will have to harvest all the sorts that can be eaten.

A mole has been busy on the lawns. Probably only a single individual, he or she has turned up about a dozen molehills of beautifully tilthed soil which I shall gratefully scoop up to mix with peat. I need to sort the greenhouse so that we can enjoy some decent autumn crops of lettuce and tomatoes.

2 September

Swirling and swooping so quickly across the track through Huppanova and dodging between the entirely unconcerned grazing cows, swallows are difficult to count. But there might have been as many as fifteen or twenty flying in front of Maidie and me this morning. They seemed not to be feeding but showing off, perhaps to each other. It is a melancholy moment when they go. The skies seem empty as summer ends, the light dims and the cold days of winter come on.

I have been steadily sawing and splitting kindling for the last few weeks and now the store near the house is full to its roof. Each day I chop ten logs from the great dump of the fallen beech tree and stack the dense, dry wood in the barn. When my daughter, Beth, and her husband, Ross, come down from Glasgow for a weekend, we will spend an afternoon and perhaps a morning filling up the stacks to see us through until next spring. It is a fulfilling task, one that brings immediate benefits, as the stoves crackle and flame and the dogs sink down on the sheepskin rugs to sleep in their glow.

But all is not yet done with this summer. This morning I noticed that the clear-felled wood to the south of the Haining Loch still lived and was fighting back. Suckers were pushing up beside the cut trunks. There were ash saplings and red chestnut, as well as many rowans. If it was left alone and deer kept out, the old wood would regenerate.

3 September

Last night when I let Lily off the lead to pee, she became very excited, spinning around at the bottom of the steps to the terrace. As the darkening nights draw in, I carry a torch and it lit a very welcome sight. A large hedgehog had wedged itself against the bottom step and was showing all of its bristles to the curious dog. It is many years since I saw Mrs Tiggywinkle scurrying up the Bottom Track with her little ones behind her, and here was her granddaughter or perhaps her grandson, one that had survived the attentions of the badgers. Last year Lindsay bought a badger-proof nesting box and hid it under the stand of Norways and ash near the farmhouse, and it seems to have become home for these eccentric little creatures.

5 September

The fencers have almost finished in the Deer Park and now I hope they will build our dog run near the farmhouse. Maidie will love the freedom and some of that aggressive energy will burn off harmlessly.

6 September

Buried deep, lost in the lush grass of the Doocot Field, Rory has found more memories of Rita and Edek's love affair, a fragment of what might have been an exchange, perhaps parting words. Ten yards from where he found the cigarette lighter engraved with their names, Rory has discovered a small enamelled army badge. Richly decorated and dominated by the eagle motif, and with *Poland* rather than *Polska* inscribed below it, it has a silver patch of solder on the back that matches a blob Rory saw on the lighter.

How and why did it become detached? If it was simply a poor piece of soldering, then it could have come off at any time. Perhaps it fell off when the lighter was lost in the field or when it was thrown away. Later ploughing could have moved it ten yards. Or maybe this keepsake was deliberately defaced, the Polish army badge torn off and cast away at the end of a love affair, perhaps when Edek was posted abroad as the Allies invaded Italy. The lighter was kept for a time because it was useful, valuable, but then the reminder of the engraved names persuaded Rita to throw that away too. Despite enquiries made amongst the older generation in Selkirk, those who were alive and aware during the Second World War, no one remembers Rita and Edek. But after seventy-five years buried in the darkness of the Doocot Field, their love affair has emerged back into the light.

Near the old western road into Selkirk, Rory's detector buzzed once more as he discovered a strange sort of medieval coin, a token of a different kind of love. Dating from the reign of Henry III of England (1216 to 1272), it had been carefully and deliberately folded in half over a thong or leather lace of some sort. Rory believes it represents a devotional act, a coin carried on a strap or perhaps around the neck. Pilgrims were always expected to give money to the keepers of shrines in return for prayers or a token that confirmed the completion of their penitential journey. Medieval Christianity was transactional. In return for money, land or privileges given to the church, benefits were expected, usually those associated with the afterlife, such as shorter terms in purgatory. It may be that this coin was folded and carried in this way to ensure it could never be used as currency by the giver until it was handed over, or perhaps blessed as a remembrance of a pious act.

Seven centuries separate these two finds but they both tell tales of love. Of course our conjectures may be fanciful in their detail, but the core of the stories are beyond dispute. Both objects are relics of love between two people and between one person and their god. These fields, a tiny geographical area, are rich; layers of intense experiences in one place.

7 September

We left the dogs and the horses and the farm to spend a day and a night in an old life. For more than twenty years we lived very happily in Edinburgh, most of that time in a small street of families whose children were all approximately the same age. Dealing with parenting, early careers, schooling, coping with all the switchbacks of constantly changing lives, five of the families became close friends and empathetic, helpful neighbours. On this day, one of the youngest of that generation of children was to be married in a ceremony at the Royal College of Physicians on Queen Street. Like many grand institutions that are less used by their membership than they used to be, they do weddings. And do them very well. I particularly liked the library and, with a glass of wine in hand, sat down next to a bookcase containing many medical texts, amongst them I noticed a two-volume history of diseases of the anus and rectum.

8 September

We spent a sleepless night trying to ignore the sounds of the city in a hotel very near to the house where we raised our children. A clear, sunny morning dawned and I found myself out early, walking down many lanes of memory: buying an armful of Sunday papers, fresh croissants and rolls, whole milk to make a decent cup of tea in the hotel room, reading the stories rather than skimming the headlines, with no dogs or horses to do.

We walked past the back of our old house, noting how the garden had, of course, been changed, and the memories flooded back. Usually I feel tears prickle at the passing of all those happy times, but this morning we just talked about what a wonderful house it had been to raise a family.

9 September

Persistent rain fell, the puddles filled and the sky was a uniform grey. I came across a new verb for such a downpour. Rain was said to be 'slenching' down. There was no sign of it stopping and I had to hang up a sodden jacket in the boiler room and change my trousers. It was good to be home. Down in the Tile Field a chorus of cows was trumpeting for some reason. They had a tremendous vocal range. One hit surprising soprano notes, another definitely a baritone. Cowsi Fan Tutti.

10 **September**

Maps matter – not only for what they include but also because of what they omit. Between 1747 and 1755, after the defeat of the Jacobite army at Culloden, William Roy was commissioned to produce a Military Survey of Scotland. So that rebels could not disappear into the trackless and unknown wastes of the more remote glens and mountains of the Scottish Highlands, good maps were essential. But Roy did much more than survey the Gàidhealtachd; he also mapped the rest of Scotland in some detail. In 1750, the first *platte*, or map, of the Haining estate and our farm was printed. There is no sign of the Henhouse, our farmhouse, and Roy's survey bears little resemblance to John Scot's estate map of 1757, drawn only seven years later.

Instead of an elaborate layout of parks and meadows, there are a few enclosed fields to the west of the loch, and around them the parallel hatching that Roy used to denote runrig, groups of open rigs and their ditches where crops were grown. Our Deer Park is enclosed by long red lines, probably representing fencing of some sort rather than the hedging around the fields by the loch. Perhaps he had seen the surviving vestiges of the medieval park pale and was recording where it ran. And even though I have often seen the patterns of runrig criss-crossing the southern flanks of the park, especially in low morning sun or under a light sugar-dusting of snow, Roy does not show it on his map of 1750. That must mean Walter is right. The Deer Park was indeed ploughed during the hungry years of the Napoleonic Wars of the early nineteenth century.

By 1843, Crawford and Brooke's map of the Scottish Borders

had reinstated the Long Track, and the woods, copses and stands of individual trees look very familiar to a modern eye, having changed little over the last one hundred and fifty years. But even though the weathered sandstone plaque that sat above the old front door of our house records a building date of 1821, the Henhouse is not plotted on this map. I wondered why it was such a secret, such a mystery.

Eight years later, Thomas Mitchell drew a charming map of the County of Selkirk that features miniature drawings of grand houses. With its colonnaded portico and what looks like a remnant of John Pringle's old house next to it, the Georgian version of the Haining dominates – even though it is in the wrong place, plotted to the south-west of the loch. Perhaps Mitchell felt his creativity cramped by the lack of space between Selkirk's cluster of houses and the north shore, and so he summarily shifted the grand house so that he could show it off better. But what seems to be in exactly the right place, to the west of the Long Track, is our farmhouse. A small, single-storey cottage, it is drawn to scale (compared with the Haining and other houses) but there is no sign of the three stone-built looseboxes that abutted the southern gable when we bought the property in 1991. While the names of Hartwoodburn and Brownmoor are written on Mitchell's map with a copperplate flourish, our little place remained once more anonymous.

Finally, in 1858, the first edition of the Ordnance Survey appeared and, despite the fact that a magnifying glass is needed, there is the Henhouse and its three looseboxes in the corner of the Top and Bottom Woods. No track appears to lead to it. All of our farm forms part of the policies of the Haining, with stands of hardwood trees in grass parks, and the Long Track looks like a very attractive avenue. But still the Henhouse is not named and, given the amount of detail on the Ordnance Survey map, that was very puzzling. It was not until the second edition of 1897 that the name is finally recorded. It all seemed very strange, but if anyone could solve the mystery of the Henhouse's anonymity, it would be Walter.

11 September

Having asked the question, I was told it would be better to meet for coffee than exchange emails. When I explained the puzzling omission of the name of my house from all and any maps until 1897, Walter answered my question with one of his own. Had I noticed anything odd about the Henhouse? Feeling like one of Socrates' acolytes, I said I did think it surprising that there were three looseboxes attached to the original cottage. What farm labourer had the means to own three horses that needed stabling? Exactly, said Walter.

He nodded for me to continue. I mentioned that on either side of the old front door there were two niches where small sculptures could have been placed, perhaps something of an echo of the Canova-style nudes that decorated the terrace at the Haining. Again, not exactly what a farm hand might have greeted when he or she returned home, weary from working in the fields.

'What about the colour of the mortar on the walls?' Walter asked. It is a deep pink, something that also surprised me. I liked how the grey of the whinstone and the pink suited each other, and when we doubled the size of the house in 1992 I asked the masons to mix new mortar of the same colour.

Walter told me in hushed tones that the reason the name of the Henhouse was not marked on maps until 1897 was because it was a nickname, and also a place that did not need to have attention drawn to it. Like many other grand houses, the Haining hosted a great deal of entertaining, as the Pringles invited house-fuls of guests from around the Borders and Edinburgh. The size of the ice house and the number of wine bins in the cellars were testament to a great deal of continuing conviviality. And because people coming from Edinburgh in particular took the best part of a day to reach the Haining, they stayed for several.

Entertainment was required, said Walter, and it led to the building of the Henhouse, with its pink mortar, sculpture by the door and its looseboxes, all at a suitable distance from the big house. Horses were kept there for only a few hours and belonged to gentlemen callers. Installed in the Henhouse were hens, 'young and accommodating ladies', who entertained the men who rode down the Long Track and passed the encouraging sculpture by the door. Watching all of this libidinous traffic, leaning on their hoes in the fields, were farm labourers, elbowing each other in the ribs, smirking or shaking their heads. It was they who dubbed our house the Henhouse, a salacious nickname that did not make it onto a respectable map until 1897. The census of 1841 shows that the cottage had become home to a farm worker. Its days as a house of ill repute seem to have been short, no more than two decades.

Well. Goodness me. All of that came as an eyebrow-raising revelation. Colourful certainly, but also surprising. Jane Austen's fan-fluttering version of life in early nineteenth-century country houses suddenly seemed a little tame. I wondered if a henhouse was thought to be as necessary as an ice house. Are there other lost henhouses with telltale looseboxes lurking down forgotten lanes? Cottages whose original, scandalous purpose has been covered over by Victorian respectability and collective amnesia? Looseboxes had led to amazing revelations about loose women. War had often marched through our farm, and now it had been replaced by horse-riding lust. An improvement of sorts.

It does not do to project values and judgements backwards, and I had to resist a mixture of Presbyterian disapproval of the antics that took place in our farmhouse two hundred years ago and a headshaking smile. The past was a very different country.

12 September

In great, glossy profusion, clutches of brambles have followed the wild raspberries around the fringes of the woods. On the same stem I found several colours: black, russet and a rich red-wine shade. If I could find the time to gather some, their tartness would make a bowl of morning yoghurt taste even better.

13 September

Soon after dawn, the flying school assembled on the east-facing roof of the farmhouse. Basking in the morning sun, the trainees shuffled along the guttering, on the ridge of the dormer windows and above the Velux rooflight, newly fledged swallows waiting for their parents to swoop in with breakfast beakfuls of flies before they take off and begin to build the skills and stamina needed for their flight across the face of the Earth. Last evening we looked up at a swallow sky, scores of them wheeling above the stables, their birthplace down the generations, their home. Perhaps they were printing its image on their memories. They will be gone soon.

14 September

Where the trees have been felled to the south of the Haining Loch, geography encourages hierarchy. Most mornings Maidie and I are out too early to see other dog walkers on the path by the loch. It runs about fifty feet below our track into the Deer Park. But at weekends we sometimes see several. Standing tall, raising herself up to her full fourteen inches, Maidie surveys these lesser creatures trotting along behind their owners. When the dogs see her and bark, she does not retaliate with the sort of frenzy reserved for rabbits around the farmhouse. Instead she offers something more lordly, proprietorial, an occasional, single bark, more punctuation than annoyance. Her obvious superiority needs no exclamation mark.

Suddenly swans lifted into the air at the farthest end of the loch, their wingtips brushing the water. Flying low, they wheeled around the remaining trees and up and over Huppanova. Until they are seen close-to, it is easy to forget how big these great white birds are.

15 September

Its footprint is much older than the fabric of the farmhouse. The date plate of 1821 seems to be definitive, but I am certain there was a cottage on the site before then, even though the maps do not record it. It may be that there was a medieval building that fell out of use, rotted down, roofless, sinking back into the ground, leaving only founds or a rickle of stones. What persuades me is not only the powerful sense of *genius loci*, the unmistakable sense of a place where people have lived and died for a time out of mind, but also some archaeology.

When we had the looseboxes demolished and re-used the whinstone to build the new part of the house, a primitive sewage system was uncovered. Little more than a chute made from a single, long slab of whinstone, it conveyed waste from pots to the outside, where the noxious heap rotted before being spread on the garden ground. Around the chute, in the south gable of the house, we found that the earth was not dark brown but black. This is the mark of an ancient midden, the repository for all sorts of other waste, organic matter (what other sort was there before the nineteenth century?) that could be spread over the kitchen garden as compost. Its position was determined by the compass direction of the prevailing wind. Because of the burn to the south at the bottom of the garden – the later conduit for sewage, until into the 1950s; when the house was last occupied, a small, brick-lined septic tank was used with an outflow directly into the burn (frogs live in it now) – the house could only easily be approached from the north. That meant the midden had to be at the south gable, out of sight and where the west wind would keep its perfume manageable. Not that people cared as much about smells of all and any sort, bad and good, as much as we do now. Sniffy is the apposite term for snooty modern attitudes.

The sun crept over the eastern horizon just at the back of seven this morning. This was a morning for Maidie and I to climb the hill in the Deer Park and check on the Old Boys, the mares and the mini-Shetlands. While the little Westie played canine tig with the minis, touching noses through the wire of the stock fence and running away, I was fascinated to watch a cloud of tiny birds, finches, I think. They seemed little bigger than butterflies and moved skittishly across the high willowherb and teazels along the banks of the Nameless Burn in the East Meadow, not staying long enough on each stem, it seemed to me, to eat the seeds. Perhaps fifty of them, they sometimes all flew in sync, swirling like shoaling herring, so that the sun sometimes caught the bright plumage of their bellies. For a moment, it glinted like blown sunlight and then they seemed to disappear as they perched once more on the stalks of the tall plants and were still. It was mesmerising.

16 September

As another day drew to its close, we sat with a glass of wine in the armchairs in the porch. Rather than switch on an electric lamp, I lit candles. They cast a much more kindly light. Lindsay and I talked of horses, hopes for the future and Grace's lessons on her pony. We watched the little pipistrelle bats feeding, their jerky, frenetic flight over the track easy to recognise.

When I took Lily out to pee at 10 p.m., she wandered off on a power sniff, weaving from side to side and further and further away. Normally she comes back immediately on the recall, but whatever she had found left no room in her head for obedience.

As my commands grew louder, I set off the Newfoundlands down at Burn Cottage, half a mile away. Their booming bark made Lily look up and she came back to me at last. Doggy stereo to the rescue.

18 September

Our fields were frosted white. Another summer passes. Returning from my walk with Maidie, I brought back unused fence posts left in the Deer Park, large offcuts and pieces of discarded rails to add to the woodpiles by the barn. A new woodburning stove has been installed. Larger, taller, jet-black and more efficient than the old one (which was also cracked), it will warm us through the coming winter. Earlier than last year, Lindsay has switched on the radiators.

Bringing in the wood, gathering winter few-ell, is an instinct as much as a task. The more I pick up, the warmer we will be in the depths of January. For ten thousand autumns, since the ice melted and the land greened, human beings have been gathering wood for their fires.

Some swallows are still here, waiting for the north wind to carry them over the Cheviots to Africa.

19 September

Too often these days I begin my morning walk much daunted. Even on the clear sunny dawn I see from the kitchen window, my problems gather like dark clouds. I seem to have endless concerns about individuals, people either not doing what they agreed to, or doing it badly, slowly, not replying to requests to get on and do it, and so on, and on, and on. Although I work alone, and prefer that, I do depend on collaboration, and other people have their own agendas, and of course their own problems. But I am not happy when they become mine. I have enough.

From hundreds, perhaps thousands, of walks, I know that the first hour of the day spent outside with Maidie will not solve everything, but eventually my heart will lift. Making myself lift up my head from the business of putting one foot in front of another, pondering what to do about some problem or other, I look at the land and its animals, switching my attention to focus on its detail, its changes. Entirely unforced, the everyday glories of what I see around me begin to soften my mood and bring some peace, or at least proportion, to the turmoil.

Instead of yesterday's frost, a heavy dew lies on the grass, the metal gates, the stock fences and the leaves. For some reason, more water vapour than usual has condensed at the end of the night, reaching what meteorologists call the dewpoint. Any lower and it would be frost. In the area of felled woodland at the Haining, what looks like a dewpond has formed. There has been no rain since last weekend, and yet it seems to be filling.

The leaves on the upper limbs of the old chestnuts in the Deer Park are turning a rich russet but one of their children, a sapling that was damaged and pushed sideways by the fencers, has put out new, pale green shoots at its top.

20 September

Last night, it was black-dark and moonless when I took the dogs out. But it was also warm. The dense clouds had wrapped us up in a giant, dark downie and I slept with little more than a sheet. I wonder if the swallows will go in this balmy spell, while plenty of insects fill the air. I saw none flying around the farmhouse this morning, but that has happened before, and then I have found they were still with us.

Tonight, it was very different: an open sky, no moon but a canopy of pin-sharp stars. I was thinking that, before Grace, the last person in my family to be born and raised on a farm was Bina, my grandmother, the only grandparent I knew. My mum's parents were Hawick mill workers and died before I was born, and my dad was illegitimate, his natural father having married someone else. That means that Grace will come to know what Bina knew, the *auld life* on the land. My granddaughter has begun riding her pony most days, bouncing up and down in the saddle, her face a picture of concentration and pleasure, pink wellies her preferred footwear. Easy with all sorts of animals, she is growing up in the clear air of the countryside but with none of the issues of remoteness that stigmatised my gran's generation, as what she called 'hicks off the headrig'. Digital technology means that Grace is aware of a much wider world.

21 September

Ten thousand September suns have shone in all their splendour on the people of this place. When the hunter-gatherers by the lochside shivered in the morning mist, blowing the embers of their fire into flame, they waited for its rays to rise over the trees of the Wildwood. High in the western hills, the legionaries on the walls of Oakwood Fort saw the sun's rim creep over the far horizon before it dipped behind Eildon Hill North and the depot below it. Travellers on the Long Track watched it rise and dapple through the autumn leaves. The sun glinted off the armour of squadrons of knights on their destriers as they rode off with Edward I to war with the rebel Scots. It has lit our history and, on mornings like these, its splendour seems eternal.

22 September

Yesterday was a day for the generations. One of my great fears is that having made this place come alive out of ruins – rebuilt the farmhouse, fenced the fields and made them productive – it will be broken up and sold when we die. My daughters have stayed in the cities, working hard, and while my son lives here with his family, his interests may lie elsewhere. But yesterday my perceptions shifted, at first a little and then more radically.

With her bombproof, endlessly patient little pony, Grace went for a riding lesson at the local equestrian centre. It was the first time she had ridden in company and at three years old she was the youngest of the group of six children. With the pony led by her grannie and shadowed by her father (I have encouraged Grace to call him her groom), she blossomed, her face a mixture of concentration and pleasure, a country child at home with horses in the hills and with others like her.

At the side of the arena, I stood with one of the dads. A plumber to trade, he had taken a couple of hours off to bring his daughter for a lesson. We talked of horses, rugby and not pushing children too hard to follow one's own interests, or what we thought might be good for them. Something to be firmly borne in mind. Grace's enthusiasms will wax and wane. She is only three. In the Borders many children, from all backgrounds and both genders, learn to ride. A powerful incentive is the annual common ridings that are held in each of the towns, hugely popular summer festivals dominated by mounted cavalcades. Many children take part and most run gymkhanas.

I took what the young plumber said to heart. It is indeed important not to push, much better to encourage. And it is an outrageous stretch at the beginning of a life that could turn in any direction, but I hope that, as a child of this place, Grace comes to love it and, in her turn, make it her home. Horses are a central part of that and the wider community. We shall see, and I shall be told to keep quiet.

23 September

Piou-piou came the distinctive call of the buzzards who nest in the Deer Park. A young one, its wingspread no wider than a large crow's, flew high over the wrack of the Top Wood, wheeling, its head tilted, searching the ground for prey. To keep the multi-plying rabbits in the cover and safety of the woods rather than coming out into the open to graze the margins of the tracks and the fields, we need more buzzards to be out hunting. But the crows mob them, especially the young ones, and drive them off. Without doubt, there is plenty of food for both species, but the crows' aggression seems a primal atavism, a need to dominate, little to do with ecological balance. The Empire of the Corvids seems to be expanding. As we sat out in yesterday evening's late sunshine, the stillness was shattered by the raucous cackle of a dense flock of crows wheeling over the stables.

On the surface of the Haining Loch, I saw a strange disturbance, something I could not at first make sense of. A bird, perhaps a duck – I was too far away to be sure – was flapping its wings frantically but not lifting into the air. Instead it was moving very quickly in a straight line across the waters of the loch, an arrow-shaped wake fanning out behind it. And then, suddenly the bird disappeared.

It was then that I understood this inexplicable drama. An ancient, prehistoric predator swims in the Haining Loch. In ferrous, cloudy water where few other species can survive, huge pike thrive, sometimes growing to three or four feet long. One of these monsters had snapped its crocodile jaws on the feet of a swimming duck and dragged it through the water. After a desperate struggle to escape, the frantic bird had been pulled down into the darkness of the loch and devoured.

Sometimes the pike themselves became prey. In 1811, a young French naval officer was taken prisoner during the Napoleonic Wars and eventually found himself and other POWs sent to Selkirk, presumably because it lay so far inland. Adelbert Doisy kept an entertaining diary:

> Some of us were passionately fond of fishing and excelled in it; the Ettrick and Tweed abounded with trout and eels of excellent quality; a lake in the neighbourhood an abundance

of very delicate pike. No one ever thought of depriving us of this agreeable pastime which proved a valuable asset for us in the culinary sphere.

24 September

In front of me in the queue at Selkirk's post office was a man who had suffered a stroke, badly impairing his speech, frustrating him until he was red in the face. Behind the counter, the young postmistress' patience seemed endless as she slowly extracted a sense of his problem (something to do with his rent book, nothing to do with the post office, I would have thought) and helped him solve it. The long exchange concluded with smiles and vigorous nodding instead of verbal thanks from the much-relieved man.

That episode was warming, moving. Perhaps because some of them knew this poor man and his struggles, no one in the long queue behind grew restive or complained. When I reached the counter, I made a comment to the postmistress about doing social work. 'Every day,' she smiled. Small communities have their curtain-twitching, knowing-everybody's-business draw-backs, but their support for the weak, infirm and even eccentric is magnificent and has long been so.

Between the ages of ten and fifteen, I had a job as a milk boy with the Co-op. Six mornings a week, I woke at 5.30 a.m. to be down at the depot by 6 a.m. Full bottles (blue for creamy and gold for more creamy – no one had ever heard of skimmed milk) were plonked on doorsteps and empties carried away to be washed and refilled for the following day.

On Saturday mornings I went with the milkman, Tommy Pontin, on the country round. In a four- to five-mile radius of Kelso, we delivered to villages and farms that did not keep cows. Many of our customers were old people and very, very few had cars. And so they gave their pension books and prescriptions to Tommy, who then went to the post office and the chemist in town, and the following Monday morning he handed over the cash and the medicine. Sometimes he would shop for specific items. I remember a set of fire-dogs from the ironmongers being stowed in the back of the milk float.

All of this was done without comment or, needless to say, any payment. Sometimes Tommy would be given jars of jam or combs of honey and once a huge leg of pork (which he divided amongst his neighbours since no one had a fridge) when a pig had been killed. It was a daily exchange of kindnesses, of inter-locking, unspoken obligation, someone who could do things for those who would have found it very difficult. Having lived to a great age, Tommy died two years ago and, not knowing, I missed his funeral. I was vexed, but apparently many people came.

Our milk float was electric, silently gliding through the sleeping streets, and all of the milk was delivered in recyclable glass bottles. We were ahead of our time, Tommy and I.

25 September

Where the crowns of the birches and geans by the Bottom Track reach over almost to make a canopy, there was a rich caramel smell. After overnight rain, the turning leaves were releasing the odour of autumn; a melancholy moment but inevitable. Further down the Long Track, rotting wood had brought forth an eccen-tricity. A strainer post had been hollowed out by the penetrating damp of scores of winters and its core had disintegrated so much that a thistle had been able to root, a proud Scottish thistle with purple flower heads lorded it above the rest.

27 September

Walter's eye for the shape of the land and what lies beneath the grass is unerring. Very assiduous in plotting his find-spots, Rory downloaded an aerial photograph of the old, sunken track I walked and the fields around it to the north of our farm, what is now Murison Hill. The Anglo-Saxon brooch fragment and the arrowhead were found close to a rectilinear feature Walter pointed to when we met. At first it looked like the characteristic playing card shape of a Roman camp, but absolutely no Roman finds had come up. Around camps there are always shards of pottery, usually coins and other material that had been lost, broken or discarded by a large force of soldiers. The Romans tended to make a mess and then leave, but there was nothing.

Walter is convinced that he has spotted the outline of an enclosed Anglian village, the precursor to the town of Selkirk. It seems to have been protected by a palisaded bank and a ditch, and had a corral for overnighting animals attached to it on the south-west side. This analysis makes every sort of sense. In the late sixth and seventh centuries, the Anglian kingdom based at Bamburgh expanded rapidly and dramatically to include the Tweed Valley. Their charismatic warrior-king Aethelfrith swept aside opposition, winning a pivotal battle at Degsastan in 603, now marked on maps as Addinston at the head of the Leader Valley, a tributary of the Tweed. Angles began to settle in a hostile community of native Old Welsh speakers, many of whom they had dispossessed. These aggressive incomers would have had to defend their gains, especially as far west as the hill country around Selkirk. The palisaded village sits on a spectacular site, looking a long way up the Ettrick and Yarrow Valleys, the direction from which trouble would come.

It seems that close to the edge of our farm more history was being made. Walter and Rory have found an outpost, a place from where cultural and linguistic change radiated.

28 September

Tinted orange by falling leaves, the puddles on the tracks are vivid splashes of colour on a grey, wet morning. Grey-bearded and red-faced, a middle-aged man has taken to running up and down the Long Track. When he reaches Windy Gates, he simply turns around and runs its length again. What a terrible way to start the day.

29 September

Over the western hills thunderheads were piling up, vast billowing pillows of cumulonimbus clouds darkening and threatening. Directly overhead, the sky was a Mediterranean blue, and there being no wind to speak of, it was difficult to tell which way the storm gathering in the west would break. Then, in a moment, something unique appeared. Over the lee side of the hills, a straight rainbow glowed. The classic colour spectrum was not curved but instead reached vertically for the border between the blue sky and the angry thunderheads. And then it came on to rain heavily.

But the Met Office forecast was optimistic. By 9 a.m. the rain would clear and the sun would shine. Fine. That was a great relief since my annual winter logging crew had assembled, lumberjacks and lumberjills, my son and daughter and son-in-law. The immense pile of logged beech from the Deer Park needed to be split and brought into the Wood Barn to dry off. Having run long leads from an outside point, the electric log-splitter (made in Canada, where they know about these things) was set up. Essentially a hydraulic ram that drives logs to a V-shaped wedge, it can ping them off like bullets if they are very dry. Two of the crew worked that, another stacked and I used the more traditional tool. My axe was made in Sweden (where they also know about these things), and its heavy head can be easily levered out if it does not go through on the first swing. By the end of the sunny and sweaty morning, we had split about three tons of beech and I reckon we still have about seven left. I will chop them serially through the winter.

My body calendar is moving from one season to the next.

About this time each year I get fatter, thicker around the middle, needing to buckle my belt one hole down. Nevertheless I still enjoyed the bacon rolls with my fellow lumberjacks and jills. The annual pattern seems to have been established to compensate for the harder physical work of the winter, wading through the mud, the wet and worse. After the year turns, I gradually shed the flab and by the time spring comes around my belt is tightened again. This up-and-down pattern used to irritate and worry me, but now I realise it is just another means of reckoning time.

October

4 October

On a dripping, damp morning the light was slow to penetrate the grey gloom. Although the temperature was nowhere near freezing, I lit the new woodburning stove for the flicker of its cheering flames. Scandinavian in design, it is a matt-black cylinder with three large glass windows so that the fire can be enjoyed from three sides. The young man who installed it schooled us in a different method of lighting the log fire. Instead of piling kindling on top of newspaper and a smelly paraffin firelighter, two logs (better if they are split logs) are placed on the bed of the firebox. Then two firelighters made from waxed shavings are placed between them and the kindling laid on top of them. The exact opposite of the usual arrangement. When the kindling is ablaze, it then falls down and fires the logs. Despite the scepticism of someone who has been lighting fires for forty years, this Scandinavian method works. Having invented the modern versions, I suspect the Swedes, Danes, Norwegians and Finns know more than we do about wood-burning stoves.

All did not begin well. For some reason, the firebox at first filled with smoke that stained the glass badly and lit poorly. The other unexpected issue is that this stove seems picky in what it will burn. Only split logs that have been seasoned undercover for a year. The old one burned anything, eventually. However, the fuel consumption of the Scandistove is much less. They recommend only one log at a time and last night it put out a prodigious amount of heat. Which is all that matters.

5 October

A gold dawn disappeared during my walk with Maidie, grey clouds from the west covering all that promise. I have seen the sunrise over Lindisfarne and I consoled myself with the memory of the rim of the fiery disc peeping over the flat horizon of the North Sea to light Cuthbert's island-cathedral. In my mind, I heard the timeless toll of the bell at St Mary's Parish Church summoning the faithful to morning communion to begin the day by taking the hand of God.

6 October

In the pool of light cast by my Anglepoise lamp, defended by a rampart of reference books, my world shrinks to pen and page. Each morning, after chores and a walk with Maidie, I sit down at my desk to write. It is not set by a window with a long, distracting view but in the midst of tables piled high with the debris of research for books published over the last twenty years. The great novelist Hilary Mantel once told me that her winning the Walter Scott Prize for Historical Fiction allowed her to buy a small flat on the coast of the English Channel and a room with a sea view where she could sit down to write. Perhaps it is different for novelists. They have to conjure worlds out of the air. I need the comfort of books around me so that I can check facts, dates, make connections, and somehow be encouraged by the presence of millions of words.

This morning my pool of yellow light is an oasis in a grey, rainswept landscape. It has been raining heavily for about thirty-six hours and our drainage is struggling to cope with the volume of water. I have already cleared one cross-drain on the Bottom Track and, since the forecast predicts no let-up, there will be more to do. But I am glad that we re-roofed most of the looseboxes last winter with zinc sheets. Unlike the green onduline, they do not look picturesque but they keep our horses dry.

Out with a reluctant Maidie – even in her wee coatie, she dislikes the rain and carefully skirts puddles and wet long grass – I saw two smart sheep sheltering by the hedge. Head-first under the leaves, they seemed content with the thick fleece of their backs and backsides being exposed to the worst of the weather. The rest of the flock was nibbling eagerly at the lush grass, stoking their systems with calories. My neighbour's drainage work during the summer has rewarded him and there are no wide ponds disfiguring the Tile Field.

7 October

Pringle Home Douglas could have been both a model for Horatio Hornblower and one of J. B. Priestley's gritty northern industrialists. He joined the Royal Navy in 1794 just as Europe was being remade, when the ferment of the French Revolution was morphing into the French Empire under the brilliant leadership of Napoleon Bonaparte. Wounded at the fierce battle of Algeciras, fought for control of the Straits of Gibraltar, Midshipman Home Douglas was promoted to Lieutenant on the night of the action as cannon fire flew around the deck of his ship. A month later, off Minorca, his brig of sixteen guns beat off a Spanish frigate with twenty-two guns. By the time Waterloo settled the Napoleonic Wars in 1814, Pringle Home Douglas had been promoted to Commander.

Retiring on his lifetime pension of half-pay, he rode up the Long Track to visit John Pringle at the Haining, his relative by marriage. Commander Home Douglas had a business proposition in mind. He wanted to build a factory two hundred yards from our farmhouse. His plan was simple, well thought out and not only would it make him wealthy, it would create many jobs.

The Tile Field would get its name from Home Douglas's enterprise. Having been drained, the bed of the old loch was found to have near-bottomless deposits of high quality clay. When Walter Elliot fenced the grass parks on either side of the Long Track in the 1960s, he took a twelve-foot rail and pushed it into the clay. Not only did it almost disappear, Walter could not pull it back up. It must still be there.

These clay deposits were at the core of Home Douglas's thinking. Dense, extensive and of very high quality, the clay beds could be scraped (which is why the Tile Field is billiard-table flat) and moulded into tiles and the ceramic drainage pipes that were needed to improve the land. The auld stane drains of the sort I broke into when my son's house was being built were very labour intensive and took a long time to complete. Clay pipes fired in a kiln could be laid much more quickly. The pleasing symmetry of Home Douglas's scheme was that all of the materials he needed were close at hand. Running on either side of the clay of the Tile Field were the Common Burn and the Hartwood Burn, ready sources of clean water for puddling the clay. A pond was formed that still survives but is now home to a pair of swans. And on the south side of the old Roman road a vast plantation had grown up on the slopes of the ridge, imaginatively known as the Big Wood. Its timber would fire the kilns needed to bake the tiles and the drainage pipes.

On yet another dripping autumn morning, I pumped up a slow puncture in a rear wheel, started my quad bike and drove off down the Long Track to see what remained of the Commander's enterprise. And the immediate answer was almost nothing, not even humps and bumps in the grass of a park where cattle were grazing. No trace could be seen of the extensive wooden buildings of the Tile Works. All I could find was the shattered debris of fired orange pipes by the gate between the two parks. And yet at least one family and possibly more people had lived there, next to this busy manufactory.

With the growth of local newspapers, and after the first census of 1841, the lives of ordinary people at last come into focus. On 31 March 1859 at the Haining Tile Works, the death of Mary Donaldson was reported. She was the wife of Alexander Wilkinson, the overseer. He wasted little time as a widower and was married to Isabella, the daughter of James Riddell, a sawmiller. On 30 January 1862, she gave birth to a son.

The baby's grandfather was a pivotal figure in this industrial story. James Riddell worked at a sawmill in the Big Wood and it supplied the timber and sawdust needed to fire Wilkinson's kilns. Both sets of buildings appear on the first edition of the Ordnance Survey that was published after 1843. The Tile Works is shown as a series of long wooden sheds where tiles dried as they were moulded or cooled after firing. There is a dwelling house with a small garden plot to the north-east, where the Wilkinson family lived. Over in the Big Wood, about four hundred yards to the south-east, the sawmill is clearly marked next to the Hartwood Burn, as it rushed downhill from the southern ridge.

When I rode my quad bike up the path that led towards its site, I found that under the willowherb and dense grass it had once been an old-fashioned metalled road like the Long Track, but not torn to pieces by the big tyres of gigantic tractors. And there it was. Completely hidden by the stands of birches and willows by the rushing burn, a rotting old wooden shed stood, filled with rubbish. It was so close to the bank of the burn that it might have housed the water-wheel that powered the belts that drove the circular saws. Uphill, the Ordnance Survey plotted three gathering ponds that helped keep the flow strong even in the dry summer months. After days of rain, the Hartwood Burn was rushing downhill at a great lick. Despite thrashing through low and wet branches, I could find no traces of the old ponds.

What I did find in my warm and dry office was a group of three photographs online. They were taken between 1914 and 1918 and two show eight men who worked at the sawmill. Two are clearly the owner or manager and the foreman. They wear ties and suits and one of them sports fancy, shiny riding boots. Perhaps he had ridden to the mill to have his picture taken. As part of the posed composition, two of the men hold very large circular saws with large, jagged teeth, and behind them are tall stacks of what might be railway sleepers. The other photograph shows a version of a small railway running across a wooden bridge over the burn. A low bogie carries a load of uncut logs.

The sawmill was still in part-time use in the 1960s, but by 1897 maps show only the dwelling house still standing at the Tile Works. It too was probably a wooden building since there are no traces of any foundations left in the grass. Pringle Home Douglas's factory and the sawmill are a pleasing example of local manufacture that accessed all of its material, energy and manpower within a radius of less than half a mile, something that should make a welcome comeback. On the wall outside the north door of our farmhouse sits a huge clay brick that was fired at the Tile Works. It is the size of a thick telephone directory, and beside it sits a length of orange drainage pipe and a blackened mass of others that were fused together in a kiln that was allowed to become too hot. Even though I never saw it, I mourn the smoke that no longer belches out of the Tile Works' chimneys. It was a sign that the land was alive, working, producing.

8 October

This morning a young kestrel flew against the southern wind, hovering over the grass parks by the Long Track before soaring. I watched it bank against the breeze, fluttering and wheeling before it flew into the blinding eye of the sun and disappeared.

Angus came to cut the grass for the last time this year. Another summer has passed but the smell of cut grass lingers around the farmhouse long enough to console me. I wondered where our swallows were. Over the Sahara now, I should think, the most testing stage of their long journey to southern Africa.

9 October

I checked on the Old Boys and the retired brood mares this morning, and they all looked well after a summer of lush, rain-fed grazing. Their gleaming condition will help get them through the winter, but Gem is now very old at thirty-one. Half Dales Pony, he is hardy like most native breeds, and despite the ticking of the clock I suspect he will still be with us next spring. The mares were lying down, their bellies the highest point, little dark bay hills.

11 October

I like my day to begin in darkness, to be up and doing before dawn, switching on lights, lighting the fire, making tea and then getting outside with the dogs so that we can all sniff the weather. The morning stars glittered in the southern sky and Orion planted his feet over the Big Wood. I imagine him standing foursquare, his hands on his hips just above his sword-belt, glowering down at us. When the pale blue glow from the east began to rise, the stars faded in moments, and by the time Maidie and I were out in the Deer Park, checking on the mares and the Old Boys, the sun was warming a glorious morning.

12 October

The colours of Scotland glowed in the sunshine. A morning wind had freshened off the western hills and, like drawing curtains, had cleared away the early clouds to let the sun splash over the farm. Blood-red berries fat with juice hung on the geans beside the old gold of their yellowing leaves. Two dense and glossy hollies mark the turn of the Top Track and, from the corner, I could see that the Corstorphine sycamore on the fringe of a far grass park had turned bright yellow. Each spring it is the first to flush with green leaves and the first to turn yellow at the end of summer. The tiny leaves of the avenue of alders by the old Roman road were flittering in the breeze of a fresh, clean, cold autumn morning.

13 October

On the night of 6 June 1841, every household in Britain was counting. Spread on the kitchen table in front of the Head of the House was a form and, with pen or pencil poised, he (almost always a he) began to write down the names, ages and professions of everyone who would sleep in the house on that summer night. A year before, parliament had enacted the Population Bill and it set out a new and comprehensive code for conducting a census every ten years. Up until that time, the pattern of calculation had been patchy since each parish did not record their numbers at the same time. There had been a high degree of double-counting and omission.

An enumerator was appointed for each census district and he had to be able to deliver all of the necessary forms and, most important, subsequently collect them all in one day, 7 June. Selkirk Parish was too large and dispersed even for an enumerator who had a horse and it seems to have been divided into eight or nine smaller districts.

Sometime in the week before 6 June, the enumerator had ridden up the Long Track, tied up his horse, taken a form from his satchel and knocked on the door of the Henhouse, moving under the ornate, semi-circular porch on its wooden columns and between the sculpture niches. It was opened not by one of a group of young ladies from Edinburgh but by the sole occupant, Anne Moscript. When she sat down at the kitchen table on the evening of the Sabbath Sunday 6 June, a date deliberately chosen when no one would be at work, Anne recorded that she was fifty-five years of age and that she was employed as a farm servant.

Almost certainly working on the Haining estate, she was probably a bondager. So-called because of the bond or contract she had agreed with the farmer or estate manager, her work was hard and unrelenting. Wearing their characteristic wide-brimmed bonnets, bodices and long, full skirts, bondagers were field workers who planted and hoed the weeds from between dreels of root crops like turnips or potatoes. They helped with the hay-making and the harvest. Around the steading, they milked cows, made butter and cheese, and looked after the hens. Bondagers did all of the manual work that did not involve horses. That was the preserve of men: ploughmen, harrowers, reapers and carters.

At fifty-five Anne was getting on for a bondager, having spent more than forty years working hard and out in all weathers, our clayish soil clinging to her boots on wet days. Her solitary life in the Henhouse suggests that it had only recently become a more respectable address. Perhaps Anne was evicted occasionally in the summer when there was a ball at the big house, an arrangement that was much simpler than moving out a family and moving the young ladies in. Moscript is an unusual name, but nineteenth-century records show several families of Moscripts living in Teviotdale. Anne may have been buried in the cemetery at Morebattle; I have not been able to find her headstone.

When we bought the half-ruined cottage in 1991, I remember the kitchen well, the place where Anne probably sat with the census form in the summer of 1841. It was a small, snug room with a long, black, wrought-iron range along the internal wall and a wide window on the eastern side to welcome the morning sun. Under the chimney we still use was the partly open fire grate that heated the ovens on either side. I still have it and the iron swee. It was a swinging arm from which a kettle could be suspended over the fire. It could be moved away from the heat so that a pot could be substituted. On the evening of 6 June there was probably evening light by the window for Anne to see what she was doing. When she noted down her age and profession, she added that she had not been born in Selkirk Parish.

Across at Hartwoodburn Farm, there were problems with spelling. In the farmhouse and the cottages by the old Roman road, twenty-four people lived and some of this community of farm workers seem to have had their names recorded phonetically. At eighty, *Isble* Cavers was a very old lady, a matriarch who had lived through both the American and French Revolutions, but her given name was surely Isabel or Isobel. Opposite Isble's name is an intriguing entry that may indicate another sort of matriarchy. Where others have their occupation listed, she was *indep*. That signified a person of independent means who did not have to work. There is no distinction made between the farmhouse and the cottages, but it is likely that Isble Cavers lived in the largest house with George and Elizebeth (sic) Riddell.

Amilia Miller and *Lillies* Laidlaw need only a little correction but what is to be made of *Chaterane* Miller, *Amilia*'s mother? Perhaps Catherine? The entry of *Nices son* opposite a five-year-old William Innis (sic) probably tells a tale of the times. He lived in the same cottage as his uncle William Laidlaw and his aunt Lillies and may well have been an illegitimate child. To lessen the stigma for the mother, little ones born out of wedlock were often sent away to be raised by relatives.

Across the flat Tile Field from Anne Moscript's solitary life in the Henhouse, Hartwoodburn was a busy community. Seven children under ten ran around the steading and the inbye fields, shrieking, shooing hens, looking for eggs amongst the ricks of the stackyard, helping bring in the milker cows in the evening, chasing rats, playing with the collies and the farm cats. Almost two centuries later, there are no children, no one living at Hartwoodburn who works on the farm. My neighbour and his colleague run both it and Middlestead at the same time, their carts and tractors often trundling along the old road between the farms.

Annie Moffat, my great-grandmother, was a bondager, and while it might have been an outdoor, even wholesome life, it was hard. I have a photograph of Annie on my desk that was taken after the Great War, when she was about the same age as Anne Moscript. The wind and rain of forty winters are etched on her lined face and her stoic, unsmiling expression, her jaw habitually set against the elements.

I write this on a damp, rainswept day with some farm chores to do before I can sit down in front of the woodburner with a glass of something warming. When Anne came back to the same front door, she will have scraped her boots and brushed the cloying mud off her skirts. In the narrow hall, her coat came off and, with her boots, was set in front of the range to dry off. And a cup of hot, maybe sweet tea grasped with both hands will have taken the edge off the cold. We have at least kept her fire burning, and not forgotten her.

14 October

A full moon sets in the west, a mist veils the Tile Field and the first hard frost of the early winter crackles underfoot. When she opened the same door, Anne Moscript saw the same sky, the same moon and stars. Having riddled the embers in the grate and set some big logs on the waking fire, she will have shivered in the unheated little cottage. Pulling on a dry coat, Anne was in the habit of going out to the old looseboxes to fill her log basket. Only when the fire in the range blazed could she make tea and perhaps some warming porridge.

Like almost all farm servants in the nineteenth century, bondagers worked for a fixed fee, cash that was paid at the end of a six-month or annual term. As important were payments in kind, what were known as gains. As part of the bargain, the farmer was bound to provide not only a cottage but also the likes of oatmeal, dreels of potatoes to be dug when needed, perhaps grazing for a cow, keep for a pig and, crucially, a supply of logs or coal, the latter much preferred. At the beginning of each term of employment, a cart would come bumping down the track and simply dump Anne's supply near the cottage door.

In the nineteenth century, coal mining was not far from the Scottish Borders. At Scremerston near Berwick-upon-Tweed and Shilbottle near Alnwick, pits supplied domestic coal. When a horse-drawn coal cart came down our street in the 1950s, it carried Shilbottle coal, its prices chalked on a slate attached to the backboard. The colliery belonged to the Co-op and costs were consequently kept down. Alarming men with black hands and faces who wore studded leather back protectors unloaded sacks of shiny black coal and shook them into our bunker at the back of the house. My gran bought big pieces of darkly lustrous cob coal we broke up with the blunt end of an axe. She also handed over a few pennies for a bag of what she called nutty slack. It was very cheap but heavy – coal dust mixed with small nuts of solid coal – and it was used to smoor a fire overnight. With the damper closed, the embers glowed, slowly burning under a thick blanket of slack, keeping the temperature above freezing. When she left her cottage to walk the Long Track to the Haining, Anne Moscript will have done the best she could manage with logs to keep her fire smouldering all day.

This morning the autumn air had a woody scent, like pencil shavings. Although I get up in the dark, like Anne, I wait for the dawn light before going out. Carrying her piece, some oatcake or bannock and perhaps some cheese for her dinner (lunch was for the quality, dinner at midday for the workers), she walked the Long Track to the big house to get her orders for the day from the estate manager or overseer. Perhaps she mucked out the stables in the morning after the Clydesdales had been led out to the fields, perhaps she helped lift late potatoes to store them in an earth and straw clamp, perhaps she helped with the morning milking or butter- and cheese-making. At fifty-five, after a lifetime in the fields, Anne's body will have creaked a little as she picked up a graip or a shovel and an indoor job will have been welcome.

In the winter months, home before dark, the fire brought to flaming life, Anne made and ate a solitary supper in the warm kitchen. She probably slept there in a box bed, and in the summer months may have moved to one of the upstairs bedrooms where it was cooler. For entertainment, Anne probably read. One of the very few fragments I have of the life of my great-grandmother Annie Moffat is that she always had a book on the go. My sister Barbara has inherited this habit, as well as Anne for her middle name. Our great-grandmother may have been a member of a subscription library in Kelso (where there were three) and so it could not have been expensive. There was also one in Selkirk, and perhaps Anne Moscript joined it. Unlike now, newspapers were not immediately discarded but passed from hand to hand and read from cover to cover. Right up until the end of her life, my grannie read every page of the *Scotsman*, including the advertisements, and especially the obituaries. But candles and paraffin for tilly lamps cost money. Winter was the season for long sleeps and, after she had smoored the fire in the grate once more, Anne went to bed early.

15 October

'Onwards and downwards!' emailed Rory Low, with news of many more coin and metal finds along the length of the Long Track and in the field around the shadow of what might be an Anglian village. He always sends photographs of the many silver pennies, and the doll-like faces of ancient kings stare at me from the screen of my laptop. Around the rims of the pennies are stamped their names and on the obverse there is almost always a cross. Secular and spiritual power preserved on a thin sliver of silver often no larger than a thumbnail.

Rory's expertise is to be marvelled at. He can tell where and often exactly when the coins were minted and one of the coins that came up last weekend was exotic, a penny from the mint of Guy de Dampierre, Count of Flanders and Marquis of Namur. Long before the euro was ever thought of, there was a currency that crossed the Channel. Rory has found several coins from the reign of Henry III that were freshly minted, prompting a perceptive observation from Walter. To prosecute his long Scottish wars, Henry's son, Edward I, was clearly emptying the royal treasury. Rory's find and Walter's reading of the land are like small pieces of an unmade jigsaw whose picture is gradually forming.

16 October

To keep the mice manageable and chase off the absurdly confident rabbits, Tabitha has at last been released from her confinement in one of the looseboxes. Now that she has been spayed and is immune from the attentions of visiting tomcats, she has taken to roaming far and wide across her new territory. The only difficulty is that it is not only her territory. The sight of her languid, feline elegance has sent Maidie into paroxysms of rage.

But the terrier's diminutive stature has supplied a partial solution to the endless barking. Adam had the idea of attaching a film that acts like frosted glass to the bottom part of the floor-to-ceiling windows that form part of two walls of the farmhouse. That successfully restricts Maidie's view outside and has been a mercy. However, a clash will come someday, but I hope by that time Tabitha is bigger, faster and a better climber.

Keeping a tight hold of the lead, scanning the foot of the hedges for a half-hidden kitten, I took Maidie up to the Top Track. Below us the shreds and tatters of a shifting mist edged over the Tile Field and the woods beyond. The dawn sun made the russet trees below Howden Hill glister like wet gold, and the hollies and geans were hung with silver, dew-drenched gossamer.

On mornings like this the landscape of precious metals shifts, seems unfamiliar, boundaries fade and the tracks disappear into the imagination. Splashes of sun light the wild cherries for a moment, and then they fade. It was as though Maidie and I walked through a forming world, emerging, morphing from centuries of poetry, of mysticism, echoes of music rising and dying away, of the distant clangour of war and the battles of half-forgotten kings.

What dunted me out of this shapeless reverie was a loud protest. My neighbour has herded some of his cows into the winter byres and they were trumpeting, their baying complaints echoing around the valley. From this summer's lush grass, he has harvested a vast store of silage, as well as some good hay, and what the cows lack in freedom will be made up by the sickly sweet flow of fodder into their mangers.

17 October

Like moths we are drawn to the light. Perhaps that is why mornings seem like the best time, a new beginning every day, a promise. This morning the light was again shifting constantly from gold dust in the west to grey overhead as the warming world woke. For a moment, the rising sun backlit the trees of Greenhill Heights, and then they vanished.

18 October

The burnt orange of the fallen birch leaves has been seeping into the puddles on the Top Track and, even on a grey day, they are luminous. In the drizzle, Wendy, her foal and their companion, the little mini-Shetland Blossom, were contentedly sheltering under the stand of willows in the centre of the home paddock. Some years ago one of the runs of fired clay drains from the Tile Works fractured and the middle of the paddock became platchy, an apt Scots word for waterlogged. Instead of the expense of repairs, we planted five willows in a small copse and fenced it. Over the years, they have drawn up most of the water and dried the ground. Now about thirty-five-feet tall, the trees supply enough of a canopy for shelter. They have grown into a giant mushroom shape because the horses often browse the lower branches. Willow contains salicylic acid, a natural painkiller and an ingredient in aspirin. The older horses know this and bite off regular doses. And, into a rich bargain, the willows are elegant and lush, the silvery undersides of their leaves contrasting with the grey-green of the upsides when a breeze blows. It is rare that any plants supply four enduring benefits.

In the morning dark, I heard a vixen shriek in the Bottom Wood behind the stables. It is too early for the mating season and so perhaps she was asserting territorial rights. But who was threatening them? When I went to check on the mares and the Old Boys, a big old dog fox loped across Huppanova no more than twenty yards from me. Not even casting a look in my direction, his tail was straight out, his head low and his self-carriage purposeful. Looking as though he was on a mission, the old fox was taking a calculated risk in the open. But why? If his libido was stirring on this damp morning, I suspected it would not impress the vixen.

19 October

So many birch leaves had fallen in the night wind that the puddles of the Top Track were like pools of blood. From the streaky colour of blood oranges, they had become the shocking, vivid scarlet of gore. Rather than a season of mellow fruitfulness, I saw the slaughter of autumn. At about eight in the morning, it came on to rain heavily and the blood was washed out.

20 October

A blearing north wind blew this morning and the leaves were falling fast and continuously. The grass by the side of the Bottom Track is a speckled yellow carpet and in a short time all will decompose and disappear into the rich earth as winter comes on.

22 October

By 1851, Anne Moscript had moved out of our farmhouse and much higher status had replaced her. The second census was more detailed, and probably more accurate. The head of each household was identified, almost always a husband, and the occupations of each occupant were clearly listed. It was a busy landscape and around the shores of the old loch, what became the Tile Field, no fewer than forty-nine people lived and worked, three generations of adults, children and grandchildren. Now there are seventeen and only my wife and I do anything like farm work. At Hartwoodburn farmhouse lived the matriarch, Agness (the spelling is still a little erratic) Dun, a seventy-year-old lady who had not appeared on the 1841 census. And at the Henhouse lived the youngest, a five-month-old baby, Barbara Spiden.

The youngest child of William and Janet Spiden, she was looked after by Agness (sic) Dinwoodie, a seventeen-year-old servant. William's occupation is listed as 'Overseer of landed estate', a slightly clunky reference to management of the Haining estate. Two other children, Martha and James, ran about the house, going in and out of the same door as my granddaughter, Grace, probably making the house come alive just as she does. These three little ones may have slept in one of the upstairs bedrooms, with their parents in the other, and Agness in a box bed in the kitchen.

Knowing a little of the lives of these cheerful ghosts animates our house; the gossamer traces of their lives are like warm ashes in the morning grate. When I riddle them and light the woodburner, smoke is drawn up the same chimney and out into the morning air, floating over the familiar fields. It is especially good to know their names and what they worked at. Work makes a place come alive.

Down by the Common Burn, at the foot of what is now our garden, one of the first things I came across was an old septic tank. It had been built from the orange bricks of the Haining Tile Works. Such was William Spiden's status as the overseer of the estate, I wondered if he had had a sewage drain dug and the tank installed. If he did, it leaked. In the small paddock between the house and the burn, now used as a dog run, the moles have dug up jet black mounds of earth, made rich by generations of sewage. Most of the outflow from six users will have gone straight into the burn. Shit history.

Old maps show a spring in the wood next to the Bottom Track and it probably supplied clean water. So many generations of tree roots have disturbed that ground, it is hard to see where a collecting tank might have been, but I have come across a scatter of more orange bricks just across our boundary. The spring still bubbles up near the Wood Barn, but we have managed to direct its hesitant flow into a drain.

Across at the Tile Works, nine people lived in Alexander Wilkinson's cottage. One of his two sons had married and given him and his wife a granddaughter, named after Mary Wilkinson. Her brother lived with them. An exotic exception, the census lists John Donaldson as a Chelsea Pensioner who had served in the Royal Artillery. The birthplace data alongside the Wilkinsons shows that this family of artisans, skilled tile and brick makers, had moved long distances to practice their specialised trade. The younger children were born near Cullen in Banffshire on the coast of the Moray Firth. Alexander and his three sons are described variously as Tilemaker, Brick and Tile Burner, and Farm Tile Work(er).

It seems that they learned their craft at Tochieneal Tileworks, where thirty to forty men were employed in a very profitable business for the farmer. His acumen in converting poor, low-yield fields of clay into large profits was much lauded and imitated at the time. It may have been the model for the Haining Tile Works. At Tochieneal, archaeologists have found many examples of fired-clay products and the D-shaped flat-bottomed field drains and the oversized bricks are exactly the same as ones I have found around the farm. When they moved two hundred miles south, the Wilkinsons clearly brought manufacturing techniques and solid knowledge of what products to turn out.

Up at Haining Rigg (sic), where a new house is now being built (very slowly), lived William Miller, a hedger. He was seventy, still working in 1851, and the only person out of a population of forty-nine who also appeared in the previous census in 1841. All of the others at Hartwoodburn and at the Henhouse are new or recent arrivals. This scale of turnover was common in nineteenth-century farms and a consequence of the feeing system. At hiring fairs in local towns, farm workers were taken on for six-month or year-long terms by farmers.

These were not slave markets, but they looked like them and could be demeaning. My own ancestors were almost all ploughmen and bondagers, and sometimes at the end of each term they came back to the hiring fair in Kelso Square, where they stood with hundreds of other farm workers and their families. If they were leaving a farm, they waited to be 'spoken to' by a different farmer looking for employees and agreed the terms of their fee, the cash payment, the tied cottage and their gains. Those who were not spoken to were left standing, publicly humiliated.

In the early 1980s, I recorded interviews with older people who could remember the hiring fairs. They ceased after the Second World War. Helen Pettigrew's family were grocers and bakers:

The hirings were busy days with all the bakers and grocers. Days before the day itself we had to get up early to start baking to have enough for everybody who wanted pies, cakes or bread.

I remember the farm workers standing in groups in the Square waiting to be hired. It was awful, very shaming for them. The first year we were up at the shop in Bridge Street, it was a very wet day and the farm workers were waiting at the door. They came to Kelso in open carts and they hadn't the oilskins then. They stood in the street knocking at the shop door to get in.

Rodger Fish's family ran a garage in Kelso and his family had relatives who came into town to be hired. His mother fed them pies and broth:

The whole thing was rather awful, it was like a cattle market, except with people. My mother used to sit at the top window in Bridge Street to watch all the carts coming past with the furniture piled up on them. This was when they were actually having to move, when they had not been 'spoken to' by their previous employer and they [hoped they] were going to another farm. They had long carts and the horses were all decorated with beautiful harness, and pom-poms and little bells on their breast collars. Invariably the fellow driving the horse and the husband would be sitting on the side of the cart. At the back there was always the horsehair settee with the wife sitting on it and all her children sitting alongside her.

23 October

As I brought in logs to light the woodburner, I thought of the Spidens, their busy little house with a baby and two young children. Where there is silence now in the early morning, there would have been bustle as Agness Dinwoodie unhooked the black kettle from the swee and poured boiling water over the oats to make porridge.

24 October

Through the morning darkness I saw Grace dancing. Framed by a lit window in her house (only a few yards from the farmhouse), my three-year-old granddaughter was giving it everything: arms flailing, feet stomping, hips swaying, head flicking from side to side. Her whole attention seemed fixed on the television in the corner of the room. Perhaps she was watching a music video. When I stepped outside with Maidie and set off the outside light sensors, Grace turned to see me and waved, her face radiant – before she carried on dancing. At 7 a.m. it was wonderful to see such uninhibited joy and those images will keep a smile on my face all day.

25 October

The first hard frost of the oncoming winter left a crust of white icing on the grass. Apparently more extremes wait, with the possibility of a very cold winter beginning early.

I have cut enough logs, the snowplough that fits onto the front of the quad bike sits right next to it and we have ten bags of winter grit. I shall check the Calor gas cylinders that run the kitchen hob and make sure the spare is replaced if it is empty.

Even though values slipped to minus four, the Old Boys, the mares and the minis seemed happy enough. It may be frosted, but there is still plenty of grass and horses will always eat that in preference to anything else.

26 October

At dusk, the sky was darkened even more when a huge flock of crows gathered over the grass parks on either side of the Long Track. Continually moving, wheeling, soaring and diving, often in spirals, they are impossible to count, but I estimated more than a thousand. Very animated, some of them even frantic, their caw-caw calls raucous, they seemed to be massing for the night roost. Now that the trees of the Top Wood are gone, save for the two old sycamores, and the wood at the foot of the Haining Loch has been felled, they perch in long, broken black rows on the electricity lines, like Morse code.

The behaviour of these descendants of the dinosaurs is strange and unsettling. I have seen crows certainly fighting, buzzing each other, even with beaks outstretched. My impression is that there are family groups (some birds are significantly smaller, almost certainly juveniles) and in a huge flock there must be rivalries over food, mates, perhaps even territory. The unrelenting, harsh soundtrack encourages the sense of conflict. Even though we often see crows pecking at roadkill (so confident that when a car approaches they hop unhurried to the verge, and then hop back once it has passed), the big flocks seem to forage in the fields, probably after worms and other invertebrates. I read that they are omnivorous. Owls are feared, and the dense flocks that mass for winter roosts probably feel safer in such numbers.

27 October

There are some concepts that bend my brain, that I simply cannot grasp. Compared with the basics of string theory or quantum physics, the clock change is a tangled knot for me. I cannot work out how the clocks going back last night is going to have any effect on my sleep, or my day. I generally get up when I wake up and even though it was nominally 5 a.m., not 6 a.m. like it had been yesterday, I still got dressed, shuffled along to the bathroom to clean my teeth and began the day. The animals don't wear watches and they needed to be fed, no matter what the law says the time is.

This biannual madness was triggered by a man called William Willett. In 1907, he wrote a pamphlet, *The Waste of Daylight*, which bemoaned the light mornings of summer. One June morning Willett was out riding his horse in Petts Wood, a suburb in south-east London, when he noticed how many houses had their blinds down. As a wealthy and successful builder whose business prospered in summer and all but shut down in the bad weather of winter, he proposed that the clocks be moved back by eighty minutes in increments of twenty over four weeks. His men could start work earlier, build faster and Willett would become even more prosperous.

No one paid much attention until the outbreak of the First World War. In 1916 both Britain and Germany adopted Willett's suggestions by turning the clocks back by an hour in spring in order to increase the production of armaments and output in other industries vital to the war effort. In the Second World War, double summertime, turning the clocks back by two hours, was adopted for the same reason.

Much later, a European Union directive insisted that all member states adopt the practice of advancing the clocks by one hour on the last Sunday in March (depending on the date of Easter, for some reason I could not discover) and turning them back an hour on the last Sunday in October. Because of its northerly latitudes, Iceland was exempted. In 2017 any member state that wished to was allowed to opt out because studies showed that the clock change had no discernible economic impact and that, in fact, it could threaten well-being. A significant number of people were found to suffer from a version of jet-lag each time the clocks were changed.

All of which just shows how utterly daft this practice is. When William Willett was out on his horse, he had presumably got out of his bed early, when it was light. When he saw all the blinds down in Petts Wood, why did he not propose the simplest solution? What he had just done himself. Get up earlier. If the sun is up at 4.30 a.m., get up. And if you do, go to bed at 9 p.m. Simple.

All of which just shows how daft chronology can be. What matters is the turn of the seasons, the light, the length of the day, the weather – and not a chronometer.

28 October

A brilliant sun rose in a cloudless, still sky and the smoke from our chimney plumed into the crystal air. Across the Tile Field, in the houses around the farm steading, more fires were lit, for it was a hard frost morning. As I walked up the Bottom Track, I thought of smoke spiralling from flames that had been kindled for ten thousand autumns in this place. By the time I reached the holly tree at the corner, I had acquired a shadow and the tissue-thin leaves of the birches had become translucent, glowing with sunlight.

Along the Top Track, Lindsay dug through the thick grass on the verge this afternoon to plant daffodil bulbs. It feels like a long haul until springtime, but we live, work and plant in hope.

29 October

Walter Elliot edited the Walter Mason Papers and, to add to what I have discovered about the story of our farm and its environs, he has been re-reading them. This morning Walter emailed to say that he had found a wanted poster from 1821 and notice of a crime committed only a few hundred yards from where I write this.

The House of Moat [this is the ruined cottage at the eastern entrance to the Howden Motte] in the Parish of and County of Selkirk was yesterday broken into while family were at Church and there were stolen Two Bank Notes of Five Pounds each with a Promissory Note by the Agent in Hawick for the British Linen Bank, dated May 1821, payable to a person or persons of the name of Shortreed for £52 and a Red Turkey Leather Pocket Book in which they were contained.

A strong suspicion lyes against William Robertson or Robinson, Carter in Selkirk (against whom a warrant is issued) who left Selkirk yesterday morning and has been traced across the County to Yarrow Water on his way to Moffat. All Constables and others are hereby requested to aid in his apprehension.

Robertson when he left Selkirk was dressed in a blue coat, yellow striped waistcoat, blue-grey pantaloons and short half boots, having iron heels. He is about 23 years of age or so, five feet five or six inches high, black hair and dark complexion.

What strikes me is the amount of cash left in a relatively humble abode – and the likelihood that William Robertson or Robinson knew it was there. The cottage was tiny and primitive, with only two rooms and no direct water supply. The other surprising element is the detailed description of the suspect, even down to the colour of his pantaloons. He sounds like a dandyish figure, certainly well dressed for a carter and not someone who was trying to move around unnoticed. There is clearly more to this tale than meets the eye. In any case, Robertson was eventually arrested, having probably evaded capture for about ten months, tried on 22 April 1822, found guilty of housebreaking – and condemned to death.

In the Borders, the condemned were held in Jedburgh Jail. Before the castle was rebuilt in 1823, the cells near the police station and the Sheriff Court House were used. A scaffold was probably set up near the site of the castle at Galahill (Gallows Hill) or perhaps near the court house and a date of 28 May set for Roberston's execution. It was an expensive business, costing £35, half of which went to Williamson, the hangman. It seems that both he and the gallows were brought from Edinburgh. There were costs for bodyguards, a man who might have been an assistant executioner, as well as the accommodation, food and ale needed to sustain them. The prisoner himself appears not to have incurred any costs.

There exists no surviving account of William Robertson's execution, but a year later Robert Scott was convicted of murdering two men after Earlston Fair and a 'frivolous quarrel'. It seems that too much ale and whisky had been taken. It was decided to hang Scott on the spot where the double murder had happened, perhaps as a deterrent, certainly to attract a crowd. Here is part of a long description of what took place:

A very strong and ponderous Gallows was made in Edinburgh which was sent out on two double carts, and erected on the very spot where the murders were committed, a few days before the execution. On the morning of Wednesday last, pursuant to his sentence, Scott was taken from the Castle of Jedburgh, and placed on a cart, with his back to the horse's tail, and the executioner facing him. The Sheriff of Roxburghshire, attended by his officers, and a strong party of the Yeomanry Cavalry of the county, escorted the melancholy procession to the limits of the shire, where the unhappy man was delivered over to the Sheriff of Berwickshire, who was similarly attended, and who conducted himself to the fatal tree which was not above a stone cast from his own house.

Soon after appearing on the Scaffold, a psalm was sung, in which Scott willingly joined, and then a most appropriate prayer was put up on his behalf, by one of the Clergy men who had so kindly attended him. He afterwards stepped forward to the railing, and, in a most earnest and impressive manner, addressed the numerous multitude of spectators assembled round the scaffold, in a short speech in which, among other things, he solemnly besought them not to allow their passions, at any time, to get the better of their reason, nor, to indulge themselves, on any account whatever, in drinking ardent spirits to excess, which always inflamed the passions, and rendered them uncontrollable; a sad example of the effects of violent and ungovernable passion, he said was now before them, in the melancholy case of him who now, for the last time, addressed them.

He concluded, by hoping that none who now heard him would ever be so cruel as cast up to any of his innocent children his crimes or his shameful end. He then shook hands with some around him, and bowed to others, then immediately mounted the fatal drop; and, while the executioner was adjusting, the rope round the beam, he was most earnest in private prayer, with his hands clasped. In a few minutes, he dropped the signal, and was instantly launched into eternity a little after 3 o'clock. After hanging about thirty-five minutes, the body was cut down, placed in the cart, covered with a sheet, and sent into Edinburgh, for Dissection.

Robert Scott was a tall athletic strong man, apparently about 44 or 45 years of age and left a wife and five children to deplore his loss.

Until I read this account, I had not realised that the condemned were accorded the dignity of deciding exactly the moment of their own death. With the noose around his neck, Scott prayed, then 'dropped the signal' and the trap opened beneath his feet. A professional hangman should have made sure that the drop would break the condemned man's neck, for if it did not, he would have spent an agonising few minutes choking to death, wriggling, kicking his legs and pissing himself. The phrase 'hangers-on' comes from those who hung on a hanging man's legs to choke him and hasten his death.

Executions were rare in the Borders, but they always attracted large crowds. Condemned to death but nevertheless enterprising, one man who met his end in Hawick had produced a pamphlet about his life and crimes, and before he was hanged he encouraged the crowd to buy copies. The proceeds would be given to his soon-to-be-orphaned children. More than five hundred were sold but the condemned man's speech and the promotion of this extraordinary piece of posthumous publishing went on so long that the Sheriff had to remind him that they didn't have all day.

30 October

Persistent mist, probably low cloud, has screened the sun so much that it is possible to look directly at it. It appears like a reflection from an old mirror, speckled and dimmed with age.

31 October

On an undated map, drawn sometime after 1856 (because it shows Selkirk railway station and it opened that year) but before 1897, the Henhouse is represented accurately and in some detail for the first time. The three stone looseboxes extend from the southern gable and are the same length as the cottage. The old semi-circular porch is marked, a garden laid out where the stables are now, and between it and the house runs a fence with a gate, presumably in case a horse got loose. A well is plotted in a small wooded area to the west. There is absolutely no sign of it now.

A path leads up where the Bottom Track runs now, but instead of turning east to link with the Long Track, it leads directly north over the ridge where the Top Wood stood and goes on to the Haining. All around is policy ground: open grass parks with stands of hardwood trees. On either side of the whole run of the Long Track runs an avenue of more hardwoods. To the east stand thirty-five trees and two dense copses, and to the west is a longer line of forty-three trees. They must have looked magnificent, especially as their autumn colours came in.

It was very cold indeed this morning, an unseasonal minus seven, and I wonder if the usual tabloid predictions of a severe winter might for once be right. It is certainly starting early and we are burning through the logs, with the new woodburner blazing all day.

The landscape was frosted white and long shreds of mist floated over the Tile Field. As the sun strengthened, rose and reached the shaded parts and the mist drifted westwards, I felt a fine, gentle spray on my face for a few moments, even though the sky was blue and clear. It must have been the morning mist dissolving.

On the Top Track, I noticed something strange, several long and straight strands of bright, white thread on the ground, criss-crossed in a geometrical pattern. I bent down to touch one and it snapped, one end suddenly curling as though there had been tension on it. They were the gossamer remains of spider-weaving, strands that had reached right across the track and between the trees and bushes. In the cold of the early morning this tracery of natural lace had become heavy with frost and then fallen. But quite why they had come down in straight, unbroken strands I don't know.

Wearing his harness, holding the pad of chalk on his breast, the ram was in with his ewes to do his annual duty. He was sniffing at one as we passed. Coy, she moved on, nibbling the grass, but did not stray too far from him. After a brief courtship and consummation, the bright pink of the chalk will be left on her back, and so it is a simple matter to tell if all the ewes in that park have been tupped. Another promise of next spring to go with the daffodils.

November

1 November

In the soft rain of a dripping morning, more scents are released as the leaves turn brown and are tinged with rotting black. Sometimes I think the earth itself is rich but its odour is difficult to describe, perhaps because it is so elemental and singular. *Humus* is the Latin word and if somehow an adjective could be fashioned from it, those syllables and overtones might be apt.

A young robin, still small and without the puffed-out, busty red breast of the adult birds, was hopping from one fence post to another. On the corner, he waited until Maidie and I had safely passed before he sang his autumn song.

The mares stood together in the rain, heads down, drookit, feeling sorry for themselves.

2 November

The spine of our house is a steep, narrow stone staircase of twelve treads that turn through one hundred and eighty degrees to reach the first floor. Beautifully made, perfectly keyed together and worn by two centuries of tired feet going upstairs to bed and coming down for breakfast in the morning, the stairs are all that is unchanged inside the house. On the top four treads are patches of red-leading ingrained in the stone. Too narrow for people to pass each other, they forced frequent waits at the top and bottom. They still do.

In the 1861 census, the Wilson family had taken the place of the Spidens. Three generations lived in the little cottage under the watchful eye of the Head of the House, Nicholas Wilson, a sixty-year-old widow and formerly a shepherd's wife. It was rare but not unknown for Nicholas to be used as a female Christian name. Perhaps, like my grannie, she sat by the fire, supervising, asking questions, passing remarks, shaking her head. Some sort of accident or debilitating illness had befallen her son, William. He was thirty-four and listed as 'Formerly Labourer (lame)'. With none of the support of the welfare state, those unable to work were sustained by their families, but William was very young to be idle. His brother, Thomas, is noted as married, but there is no record of his wife living in the cottage. Had she died, he would probably have been marked as a widower. Thomas's daughter had been given her grandmother's name of Nicholas, perhaps a family name, and her brother is named after his father. The status and whereabouts of the mother of these two children is a mystery.

Their aunt, Elizabeth, is said to have been an 'Assistant at Home'. It may be that she looked after both her mother and her lame brother. Thomas was the only member of the Wilson family to be working. He was an 'Ag Lab', an agricultural labourer, on the Haining estate's farms. The impression of a household struggling to make ends meet is reinforced by the presence of George Hall. He was a lodger at the Henhouse, like Thomas, an agricultural labourer, and no doubt a welcome source of income.

For the first time, the census counted the number of bedrooms in each house. There were three in the little cottage. When we first saw it, albeit in a ruinous state, with the roof partly collapsed, there were only two small bedrooms on the first floor. What had been used as a sitting room might have been converted into a bedroom for George Hall. That will have made the kitchen the only communal space and, with seven people, it will have been crowded.

Up at Haining Rig, William Miller, the hedger who lived there in 1851, had since died but his widow, Catherine, still lived in the tiny cottage. It was another very crowded place to live. Twenty-five years ago I first saw the ghost of Haining Rig, lost in the midst of a dense forest of sitka spruce that had been planted all around it. It was as though the wood had consumed the house. Self-seeded trees grew up through the floors and had pushed over parts of the walls, bullying the cottage out of existence. It had a small porch, essential for leaving wet clothes and boots, and two very small rooms. After the years of assault from the growing sitkas, it was difficult to tell if there had been a first floor. Perhaps there was a low attic. It was a sad, damp, abandoned place, the cheering fires in its broken hearth long since gone cold.

At midnight on 7 April 1861, when the census was counted, seventy-seven-year-old Catherine Millar occupied one bedroom at Haining Rig and crammed into the other were three people. Betty Loch was another widow, this time of a shepherd, Catherine Hardie was a visitor and thirteen-year-old John Marrs is described as a boarder.

A pattern familiar in Scottish demographics is already clear: women often lived longer than men.

3 November

Last night the acrid stink of terror filled the air. In the East Meadow the two mares fled to the farthest corner of their field, and after searching for them by torchlight Lindsay found them standing close together shivering, very frightened. In the darkness we listened for the hoofbeats of the Old Boys, bolting in panic and fear. But they have a seven-acre paddock with dips and swales where they can hide themselves. Down at the stable yard we had brought in all of the other horses and played the radio loud to drown the detonations and screeches.

It was Fireworks Night and the sky above the town was filled with rockets and explosions of all sorts. It meant that eight of our horses were directly exposed to twenty minutes of bewildering, deafening terror. The ear-splitting bangs echoed off the wood behind them, on the south side of the East Meadow, so that it sounded as though explosions were bursting all around the horses, leaving them nowhere to run. Flight animals, it is their primary instinct. On either side was more alarm. For some reason, the firing of the rockets made our electric fencing ping and squeak as though a powerful current was passing through the top wire.

It was misty and our torches could not penetrate far into the darkness. We had to stop and listen for the animals panicking, for snorts, whinnies and the drumming of hoofbeats, not easy with whizz bangs every few seconds. Our worry was that they would jump out of their paddocks or barge through the fencing and injure themselves, and how would we retrieve them in the darkness if they galloped down the Long Track towards the roads and away from the explosions? But we heard little, the cloying mist dulling sound as well as vision.

This morning I went out at first light to check on the minis, the mares and the Old Boys. All were upstanding, some even grazing in the rain. I realise that people like fireworks but animals most certainly do not, and surely some consideration must be given.

4 November

It turns out that we breed butterflies. Last week eight red admirals and painted ladies fluttered around the house. Thinking that they had somehow managed to get in through an open door or window, we carefully captured them all in cupped hands and let them fly outside. But it is so late in the year for them to thrive, too cold and wet, that Lindsay realised that they must be breeding in the warmth of the farmhouse – amongst my tomato plants in the conservatory. When I watered them yesterday, I noticed that some of the leaves had been nibbled, but I could not find any caterpillars. There were plenty of small yellow flowers to supply nectar.

The tomato crop has been modest and I had planned to uproot all of the plants and just grow lettuce. We eat plenty. But I will leave them a little longer to see if any more of these most beautiful of insects, the ghosts of a dead summer, flutter around the winter fires. Butterflies at Christmas would be strange but magical, following us like Philip Pullman's daemons.

5 November

Moving slowly, almost imperceptibly, from different parts of the grass park by the Long Track, all at the same deliberate pace, in the same direction, the sheep flowed together like a river-flock. Smaller tributary groups merged into one as though guided by ghostly shepherds and collies. It was a hypnotic, mysterious slow-motion procession in the half-lit drizzle. After a few minutes those who led stopped and began grazing.

By late morning, the sky had brightened and the drifts of yellow birch leaves across the track glowed like a fallen sun.

6 November

It was far too cold this morning, far too wet yesterday, and for much of the summer there was far too much long grass on some of the tracks. Maidie has strong opinions on what is good weather for a walk. This morning she planted her paws at the bottom of the Bottom Track and refused to budge. She may not like the cold, the wet or the long grass, but she would eat the cat.

The lush grass was heavily frosted, like luxuriant hair suddenly gone grey, but it will soon be grazed down and we will need the ring feeders filled with haylage soon for all those who are permanently outside. Winter is an expensive time with horses – and when vets are more likely to be called. I wonder if all of the Old Boys will get through a hard winter. I will check them every morning now.

On the Haining Loch, two swans swam in the grey morning and a light went on at the big house. I used to imagine it being taken over by homeless squatters (why not?) but I realised this morning that it was an outside light with a motion sensor. Perhaps a passing animal had set it off?

7 November

Even though the yellow and russet leaves still cling to the trees and their colour lights the grey morning, winter is here. After three days of rain, some of it very heavy, the land is saturated, the burns swollen, and all is mud and slippery clatch. In front of us, we have five months of wet coats drying in the boiler room, taking out feed to the horses in the outbye fields, and endless days of dreich and gloom when the sky seems no higher than the roof of the pick-up. But the evenings can lift the days, the hypnotic, warming, dancing flames of the woodburner, a glass of something good and plenty of hot, solid, filling food. The winter insulation will be shed in spring.

8 November

Friendships come and go. Even the most intense usually flower and then fade because circumstances change, locations shift and interests diverge. From the gregarious social ferment that university can be, when I seemed to know scores of people very well, to my life now, the difference is stark. I have five people I would call good friends, but I see them rarely, at most every couple of months. That is partly a function of living on a little farm where the nearest neighbour is half a mile away and the nearest city forty-five miles to the north.

This coming weekend is the exception to all that drifting apart. At lunchtime, we will drive up to a stunning modern house perched high on a river cliff over the Tay about ten miles north of Perth. We will meet our old neighbours from Edinburgh, 'the street'. On Friday nights, after a week of work, we would often gather in our kitchen. Once I remember coming back from Barra in the Hebrides with a sack of cooked langoustines and lobster that needed to be eaten immediately. It was chardonnay then rather than sauvignon blanc.

When we moved down here to the farm, our old friendships might have withered into exchanges of Christmas cards. But such was the strength of those early bonds we decided to come together once a year for a weekend of three Friday nights in this beautiful house. Even though we have all grown old together, the years seem to fall away, at least for the first Friday night. The future may well not be what it was and that space is mostly occupied by children and grandchildren. But the past often comes alive again, especially after a glass or two.

What particularly attracts me is the shape of each of the two full days. Chronically unable to lie in, I get up and walk into the village for the papers at 6 a.m. After tea, toast and an hour with the news, I make a fried breakfast on a production line, as our friends pad downstairs one by one. I like the early time to myself, looking out over the great river, reading and looking forward to the warmth of good company.

There used to be six couples, but, very sadly, one husband died in his early sixties and now we are eleven. Departures will begin to happen again, but for a time the street will meet once a year. You don't make old friends.

12 November

Home to familiar sounds and smells: the purr of the shepherd's quad bike up at Brownmoor, the trumpeting of the cows in their winter byre at Hartwoodburn and the squelch of the clatch as we bring in the horses off the sodden fields. All of it a comfort.

13 November

At first light, the partridge family were skittering about the home paddock. Spooked at the sight of a little white dog, they scattered in all directions. But Maidie had more interest in sniffing after the trails of animals that had passed close to the house during the night.

It was a moonlit dawn, the light black-and-white in the west, like an old film, warming to a pale yellow in the east as one planet rose and a star set. For a time both were visible in the morning sky, the moon big as it slipped behind Peat Law and the sun brilliant over Greenhill Heights. A hard frost for the Old Boys, the minis and the mares in the outbye fields, the best sort of winter weather for horses. The worst is cold, wind-driven rain and no doubt we will see plenty of that before the clocks change at the end of March.

14 November

The last of the night-snow speckled surfaces that were raised off the warming ground. White flakes lay on the compost in plant pots, on the ash bucket and on the seats of the benches down at the stable yard. The first snow of the winter, it was strewn lightly on the western hills, the line of the track up Newark Hill still visible under what had been only a sugar dusting. The morning sun will make this harbinger of what I fear will be a cold winter disappear, but on the twigs and branches of the naked trees hung a tracery of frozen water droplets as though caught in the act of dripping. Stars of ice reached across the puddles like Christmas tree decorations and the old spring at the foot of the Bottom Wood glistened solid and motionless.

Before more snow comes, now is the time to make preparations. I have ten bags of road salt for the tracks but I need the winter grit mixture of sand, salt and fine particles of crushed stone. It lasts and needs only to be spread every few days, but I cannot find a local supplier. The other priority is to bring forward the big logs for our second, much larger woodburner and I will use a barrel-sized basket outside the porch, cover it with a horse rug and weight it against the wind.

This morning's sun will warm the house through its many windows and melt the tissues of ice on the tracks, but when the bad weather comes and walking is precarious, everything essential needs to be readily to hand.

Our absolute attitudes to warmth in the winter have changed during my own lifetime. All that will have heated the Spidens, Ann Moscript and the Wilsons in the nineteenth-century Henhouse was the range in the kitchen and perhaps the open fire in the other room. A hundred years later, the sole source of heating in the council house I was raised in was a coal fire in the sitting room. On bitterly cold nights when ice formed on the inside of the panes of the bedroom windows, I slept under thick blankets, a quilt and sometimes an overcoat on top of all that. What kept me warm was my own body heat. Sometimes the tip of my nose became cold. When the alarm went and my bare feet touched the linoleum, it was like stepping onto ice, and we all grabbed our clothes and rushed downstairs to the sitting room where my mum had lit the coal fire. That was where we pulled on socks, warm clothes and sat with a steaming bowl of porridge. Only then, fortified and clad, could the day uncurl and begin. Such is early conditioning, I still cannot sleep in a warm bedroom. In summer, the window has to be open, and in winter the radiator is kept just above freezing.

15 November

I met an old friend in Selkirk this afternoon and he amazed me. Rolling back the decades, he told me that tomorrow, Saturday 16 November, will be exactly fifty years since we first played rugby against each other. Since the tender age of seventeen (and I was tender when the final whistle went), I had appeared in the Kelso front row and my friend was also a young starter. Colin and I only ever talk when we bump into each other by chance, but the bond of old-fashioned comradeship is always remembered with a smile and a handshake.

Like the common ridings, the summer festivals where horses are everywhere, rugby is woven into the fabric of community life in the Borders. Each of the main towns has a club and a surprisingly large stadium, out of scale with the population size. The Border League was the first rugby competition in the world, outraging the blazer-wearing alumni of posh private schools and the ancient universities who could afford to cling to amateur ideals and who ran the Scottish Rugby Union. In the Borders, working men played the game, butchers, bakers and bricklayers, and the competition between the towns was fierce, forging generations of excellence. And because rugby was preferred to football by small children playing in parks, their catching and passing skills were ingrained from a very early age. The likes of Hawick, Gala, Melrose, Kelso, Jedforest, Selkirk and Langholm regularly beat the old boy clubs of Edinburgh and Glasgow and yet it took many decades for their players to break through the snobbery barrier and get into the Scotland team, controlled as it was by the blazerati.

I loved the spirit of the amateur game, the fun as well as the glory. In 1958 Ian 'Basher' Hastie, a gifted Kelso player, was at last selected to play for Scotland against France. We watched the match on the grainy black-and-white TV of a neighbour and erupted when Ian scored a try in the corner. I was mystified when my grannie shook her head. On the following Monday morning, Basher was back at work at Kelso railway station and when he cycled past our house my gran waved her stick at him. 'Here! Come here you!' When this huge man dutifully stopped, he heard her complain, 'Why did you not run under the posts to make the conversion a bit easier?' The huge man's face fell. 'Sorry. Sorry about that, Mistress Moffat.'

Rugby is now a professional sport and it has moved far from the game I used to play. Heroes no longer walk, or cycle, down the streets of Border towns. Like professional footballers, they inhabit another world. But I smile at the memories shared with my old adversary, Colin. And the thrill of running onto a pitch to be watched by thousands is a moment and an image I will never lose.

16 November

We returned late and in darkness from a long, three hundred and seventy-mile round trip to Inverurie on more equestrian business. From the Borders to the Don Valley in Aberdeenshire, it is a journey up much of the length of eastern Scotland that moves through our culture and history. Crossing a range of southern hills, the Firth of Forth opened below us, and after the hills of West Fife and Kinross we crossed the Tay and on through the undulating, well-drained and fertile fields north of Dundee. On this long road, it is possible to observe the wash of centuries of change across the landscape without leaving the car.

Road signs are the markers of the movement of peoples. South of the Lammermuir Hills, the Lambs' Moors, the Dark Ages kingdom of Northumbria is remembered in the predominantly English names like Selkirk, Galashiels, Stow, Middleton and many others. The earlier stratum of Old Welsh sometimes pokes through in the likes of Peebles (from *pybyll*, for a shelter or a shieling) or Penicuik (the hill of the cuckoos), and we pass by the conundrum of Edinburgh. Who was the Gaelic or Old Welsh-speaking *Eidyn* and why was his name attached to the Anglo-Saxon *burh* for a fortification?

In Fife, the mixture of recent English, earlier Anglian, Gaelic and Pictish reflects a jigsaw pattern of ownership and multilingualism in North Queensferry, Crossgates (a crossroads), Kinross (Gaelic for the 'head of the promontory' that stretched into Loch Leven); names like Pitreavie and Pitlessie have the Pictish prefix for a parcel of land.

Beyond Dundee, the road signs show extraordinary eccentricity, place names unlike any others in Scotland, probably a cocktail of Pictish, Norse and Gaelic. Happas, Memus, Bogardo, Idvies, Bogindollo, Ballindollo, Edzell and others. The 'dollo' suffix is probably the genitive case of the Norse *dalr* or perhaps Old Welsh, *dol* or maybe Gaelic, *dail*. All mean a valley. Edzell may be a very heavily corrupted version of a name that incorporates one of these. *Baile* in Gaelic is a settlement and that may make Ballindollo something like 'the settlement or farm in the valley'. Happas, Memus and Idvies will forever remain a source of wonder and delight.

17 November

More and more rain saturates the fields and turns the tracks into torrents. In the last few days there has been extreme flooding in South Yorkshire and Gloucestershire. This is fast becoming part of a new pattern, the effect of more extreme weather in turn caused by climate change. Politicians will be forced to react to these symptoms (to say nothing of the financial pressure insurance companies will exert) by building flood defences. But my hope is that the cause – the undoubted fact that our planet is burning – may be recognised by more and more people. Perhaps they will force our boneheaded governments and those in other countries to react before it is too late. If there are votes in tackling global heating, then politicians will deal with it. I hope.

19 November

During the night the temperature dropped to minus ten, the coldest so far this winter. At 10 a.m., in hazy sunshine, it was still minus four and I loaded the woodburners with logs. Both are blazing. The still morning sparkled with frost that did not thaw, the lacy tracery of the long grass very delicate. The metal gates were so cold my fingers stuck to them for a painful moment. I need to wear gloves, but the fiddle of taking them off to unpick a knot or make a note is maddening. Huddling close to the hedges and woods on the edges of their parks, the sheep were seeking places where the temperature might have been a degree or two higher. This deep cold in the early winter will see some of the older ewes die if it continues, and especially if snow comes. Fingers of ice were reaching across the Haining Loch.

20 November

We are ice-bound. During the night, more ice crept across the tracks like freezing lava. Probably because the temperature rose slightly, more water trickled out of the Top Wood and, with no trees to soak up moisture, it oozed from the warmer grass to the exposed surfaces of the tracks and froze, making movement of wheel and foot difficult and even dangerous. The good news is that I have found a supplier of winter grit and twenty-five bags will arrive, but not until next week. If we run out of grit, no vehicle can go anywhere and we will tread with great caution. Lindsay attaches a wire version of climbers' crampons called Yaktrax to her boots and I have a pair somewhere that will come into service soon. Already, in late November, we find ourselves in the deeps of the winter.

21 November

The bright pink crayon marks on the backs of the ewes catch the eye on a grey morning. They are graphic evidence that the rams have done their duty and the tupping is complete for another year. An ancient seasonal cycle sees the sheep moved to better grass in one of the western parks, and as the days grow dreich and short the work on my neighbour's farm settles into a rhythm of feeding animals. The grass will soon turn bitter and feed will need to be taken out to the inbye fields to keep the pregnant ewes thriving through the winter to come. Our own winter cycle is also beginning; the first round bale has gone into the ring feeder in the East Meadow so that the Old Boys can continue to do well, rugged up, warm and contentedly munching.

Sheep rearing in the upper valleys and on the hills that shelter them can be very different, the weather often significantly worse, more snow and ice making moving the flocks more difficult. Walter Elliot once took me out to Ettrickbridgend to meet Tommie Wilson, one of the last of the hill shepherds. Now long retired, he remembered summering out with the ewes and their growing lambs up on the high pasture, sleeping out at the shielings with only his dogs for company. And he recalled some of the winters with a shake of the head. Drifting snow was the great killer and many times Tommie used his crook to prod for trapped ewes. He told us a tale of a heavy fall of snow very late in the winter, just as the hill ewes were beginning to lamb. One was found in a drift that had overwhelmed one of the dykes that run across the high pasture. In a snow pocket on the lee side, the ewe had given birth and in that white space had begun to suckle her lamb.

When I told Tommie we were at the Henhouse, he said, 'I mind fine driving sheep along the old road below you. I was taking them from up Ettrick across to the mart at St Boswells.'

In the 1950s and early 60s, there was much less traffic and not many hill farms had access to the big transporters that carry beasts to market now. So Tommie and his four dogs drove flocks at the end of every summer along the side roads to the mart and the railway station beside it. 'If a car did come, the collies saw it first and just set themselves in the middle of the road and wouldn't move until it stopped. They sort of faced it down.'

Walter was born and raised in the valleys and I enjoyed listening as he and Tommie reminisced about country dances at the Boston Hall, the shepherds' meets at the end of the year, and at the beginning. Any sheep that had strayed onto another hill farm were brought to these gatherings to be claimed, as was whisky, food, music, song and storytelling. Rare and very sociable, these occasions were highlights in a solitary life of self-reliance and reflection amongst the hills and the high country.

22 November

Like the clacking of a football rattle, the cackling of the crows builds to a crescendo in the hour before dawn, as they lift invisibly into the dark sky. Only the receding sound tells me they have taken flight. An increasingly large flock of perhaps one or two thousand has taken to roosting in the fields around the farmhouse and when they rouse themselves the noise builds surprisingly. If it was not so familiar, the cacophony might seem sinister. It is definitely not birdsong.

24 November

So that the working days do not shrink to only the six or seven hours of light available, I start in the dark. After all the dogs are done, fires lit, my tedious but necessary routine of exercises complete, and emails done, Maidie and I walk out into the dark. Being a white-haired terrier, at least I can see her. I might fall over a black lab. This morning we heard but could not see another dog walker on Huppanova. But when he or she moved to the near horizon there was just enough light to make out a dark silhouette.

In the distance, against a black sky with no horizon, I could see the brilliant overhead lights for a new set of roadworks on the main road. Traffic lights added a little colour as they blinked in the dark, sleeping land.

25 November

I seem to be travelling too much, and I long for the peaceful and productive routines of the farm. Taking the train to Fife for a meeting, I crossed the Forth Bridge. Normally, the spectacular sweep of this stunning structure never ceases to impress me, but there was so much mist billowing up the Firth that there was little to see.

Driving home up the Long Track in the dark, I turned a sharp left at Windy Gates and the headlights flushed a young fox from the dieback of the Top Wood. Not much bigger than a cat, it loped along in front of me, not weaving from side to side or diving back into cover but showing all the confidence of a top predator. Later I came across what might have been one of its kills when I took Lillie out to pee. A big rabbit, not long dead, lay behind the box hedge below the house. I had to fight with the dog to get it out of her mouth and fling it over the hedge. Where the young fox will find it.

26 November

I watched two buzzards glide over the Young Wood south of the East Meadow. They were set upon by a gaggle of angry crows and driven off towards the high ground of the Deer Park. Was it territoriality? The buzzards surely presented no threat. There was no food to fight over and it seemed to me to be nothing more than a dispute about who owned the airspace.

27 November

In this landscape, time races back and forth across the centuries. Last weekend Rory Low found a beautifully decorated Anglo-Saxon stud that perhaps formed part of a shield, or more prosaically a piece of furniture. It turned up in the field beyond the woodland strip that borders the Doocot Field. More and more evidence of an Anglian settlement, the place that probably gave Selkirk its name (*Sele* in Old English is a hall and *circe* is a church. In 1124, the name was recorded as *Selechirche*), is coming up out of the ground. Rory's stud hints at richness, as does the brooch he found. Perhaps they belonged to the Anglian lord who built the hall and gave money and land to the church.

Nearby, he has found relics of war: two belt buckles worn by soldiers in the War of the Three Kingdoms. They were almost certainly lost at the bloody slaughter that took place at Philiphaugh in 1645 and were probably dropped by two of General Leslie's Covenanting cavalrymen.

Two and a half centuries later, the landscape these troopers rode through was changing once more. By 1899 and the publication of the new Ordnance Survey, the tile works and the house that lay to the north of the long sheds had completely disappeared into the long grass of a park, what is shown as open farmland. But elsewhere there was bustle and activity, and according to the 1891 census almost sixty people lived and worked in and around Hartwoodburn Farm. It was the zenith of the age of high farming. Britain's farms not only had a domestic market to feed, with railways reaching into the heart of the countryside and distributing animals, grain, cheese, butter, bacon and all the other goodness that grew out of the land, but they also supplied the Empire, its armies, the Royal Navy and the civil service in India and elsewhere.

The texture of everyday life on these highly productive farms lies just within my own extended memory. In 1890 at Cliftonhill Farm near the village of Ednam and the town of Kelso where I was raised, my grannie Bina was born into the *auld life* and all of its richness and rhythms, daily and seasonal. From the time she could walk steadily, Bina took a basket to search the steading for hens' eggs, went ratting in the stackyard with the terriers, and with a small stick and a loud voice helped herd the cows in for the evening milking.

For six years, she lived at Cliftonhill, one of them spent walking down the hill into Ednam to the little primary school. Bina's grandfather was first horseman, head ploughman, and she remembered him grooming the huge Clydesdales that pulled the swing plough through the heavy, fertile soil of the fields by the Eden Water. In the evenings, the work done, Bina sat with him on the corn kist, the secure chest where horse feed was kept out of the reach of the rats. Like many men, William Moffat smoked a pipe and I can remember Bina showing it to me. She had kept it after his death in 1897. On top of the bowl, it had a hinged silver cap with small holes in it, like a salt cellar, to slow down the burn of expensive tobacco. Very sadly, the old pipe was lost, perhaps thrown out when Bina died in 1971. All I have of my great-great-grandfather, the first horseman, is a small pale white and black box made out of ram's horn. Beautifully carved, it closes tight and I wonder if William kept his matches in it.

29 November

Even though a cold, pale light was rising in the east, the dawn stars, the brightest – Sirius the Dog Star, Arcturus, Alpha Centauri – were still visible. Clear and still, the morning was spared the cackle of the crows and only a solitary buzzard's *piou-piou* echoed across the grass parks. At 6 a.m. the sun was well below the horizon, but it backlit the bare trees on Greenhill Heights, even though it would be another hour and a half until its rays reached us. At minus five, there was a gossamer film of ice on the tracks, but I was sure it would shift as the morning warmed. Out on the Long Track, I watched the land change from a cold grey to blue and then pale yellow as the light strengthened. Even though it was cold, I walked slowly, letting Maidie sniff for as long as she wanted in the dieback. The colours are delicate, subtle, fleeting.

30 November

The world is white. No snow has fallen, but the hardest frost of the winter gripped the land last night. Twelve degrees below zero has frozen the puddles and the water troughs in the fields solid. The Old Boys can drink from the free flow of the Nameless Burn, but we will probably need to break the ice in the troughs that water the mares, the minis and the others who will go out later.

Maidie led me up a rabbit track through the dieback in the Top Wood and we stood for a few shivering, precious moments at the summit of the ridge as the white land glinted below us. In the dawn light, our panorama was graphic, from the hill and hummocks of the Deer Park to the east, around to Greenhill Heights with the unrisen sun glowing yellow behind, to the tall TV mast due south and its five red warning lights, warm and beautiful, then to the west our little valley unfolded and rose up the motte and the hills of the Ettrick Forest beyond. Completing the winter landscape was Newark Hill and Peat Law to the north.

Wingbeats whooshing, necks outshot spear-straight, two swans wheeled below us and flew over the Haining Loch.

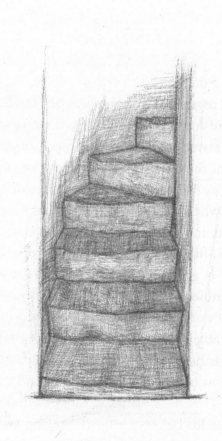

December

1 December

Even though she died almost fifty years ago, my grannie Bina and I sat down together to have breakfast this morning. Last night Maggie and Archie Stewart came to a talk I gave and they had a gift that moved me very much. They farm at Cliftonhill now and brought me a bag of porridge oats grown in the fields once ploughed by William Moffat, where Annie Moffat hoed the weeds and where Bina helped at harvest time. These are the fields of memory for my family, a landscape of loss and renewal, sodden with winter rains and then bright with ripening corn. Their soil is grained into my hands. Part of the harvest was given to families in the cottage row as their 'gains', and so when the pan of oats began to seeth and steam I took my bowl to the kitchen table and sat down where Bina was waiting for me, her words echoing across the years. 'Get something hot and substantial into you first thing,' she said to me – every morning.

Some of the fields below the farm steading border the Eden Water and I noticed that the porridge oats were branded as produce from the Eden Valley. When I am going that way, I always stop by a field entry at Cliftonhill to look out over the southern vistas that Bina knew so well. They are unchanged and, to my eye, Edenic.

The prodigious harvests of the golden age of high farming were increased even more by benign climate change in the middle of the nineteenth century. Between *c.*1300 and 1850 the global climate became very cold, frosty and wet in a series of long periods that became known as the Little Ice Age. The first recorded Frost Fair was held in 1608 on the thick ice that covered the River Thames and the last took place in 1814. Pieter Bruegel the Elder's famous painting *The Hunters in the Snow* showed a scene common across Western Europe in those wintry centuries, and the most apt image for Scotland is Henry Raeburn's portrait of the Reverend Robert Walker skating on Duddingston Loch in Edinburgh in the 1790s. Along with curling, it was a winter pastime that became very popular, since in the colder weather rivers, ponds and lochs could be relied upon to freeze regularly.

On the Ordnance Survey map made after 1856 for our farm, there are three strange rectilinear shapes that seem to have no agricultural purpose. They look like ponds that were dug out to gather water from the Nameless Burn that runs at the foot of the East Meadow. Under the thick dieback, I was able to make out at least two embanked lines of mounded earth when I went looking with Maidie this morning. I am certain that the three areas were allowed to fill with water so that they formed a skating and curling rink for the Pringles and their children from the big house. On the plan made for John Pringle in 1757, they are marked as a long pond, perfect for creating alternate ends for teams of curlers. The Nameless Burn now bypasses these depressions, but it would be a simple operation to dam it and divert the flow. But I suspect we have enough ice to deal with.

2 December

When the nights began to close in and the first flurries of snow fell, Bina used to stare out of the sitting-room window and talk of 'drifty days'. It was a phrase from the *auld life* on farm places. As its name suggests, Cliftonhill is perched on a ridge above the

Eden Water and the village of Ednam. Exposed to the snow, those who lived there in the 1890s must have thought that the time of severe winters had come back. In 1891 the first use of the word 'blizzard' was recorded, and it appears to derive from the German *blitzartig*, a storm that comes like lightning. When Bina was three years old in the winter of 1893–4, a great deal of snow fell and many mature trees were blown down across the Border country as the blizzards blew. The following winter saw unrelenting frost from 30 December to 5 March 1895 and many days of heavy snowfall. A year later, the blizzards came roaring back. Like all who worked on the land, Bina feared the extremes of the winter and prolonged periods of snow are what planted the image of drifty days indelibly in her memory. There were long days when she sat staring out of the cottage window at the still, white landscape where all that moved were hopeful robins hopping around the branches of bushes, looking for frozen berries.

When the land was hard and frozen with deep snow, Cliftonhill simply survived, as pathways were shovelled clear, ash from old fires thrown down as winter grit and hammers taken to smash open the solid water troughs. The plough horses, the great Clydesdales, needed to be led out of their looseboxes and even if the fields were deep, it was better that these equine engines kept their muscles moving, if only for an hour or two. As time's wheel turns, we find ourselves doing similar things, as winter closes in around us and the drifty days are not far off.

3 December

A smoky golden dawn brought milder weather, a welcome break from ice and a brisk wind to dry the sodden fields. This morning's sun picked out the sheep my neighbour has put into the Deer Park and it is good to see that herb-rich grass being eaten down, bitter though it will be.

4 December

The 1891 census listed ten people packed into the Henhouse. First was the Head of the House, Andrew Harvey, a ploughman, and with his wife, Catherine, he had six children in the tiny cottage to feed and clothe. Noted as being between the ages of twelve and one, five had been born at regular two-year intervals, with the youngest, James, perhaps an afterthought. He was three years younger than his brother, George.

In an age before reliable contraception, this intense pattern of procreation was not uncommon. Most women are unable to conceive while they are breastfeeding and it looks as though Catherine fell pregnant as soon as she had weaned each successive baby. In such crowded conditions, life must have been a series of never-ending demands on her time and attention: feeding, potty-training, endless washing of clothes and keeping an eye on toddlers in such a tiny space. But in the late nineteenth century farm workers were encouraged to have large families. More hands could help at harvest, in the stackyard, with hens, gathering the wild harvests of berries, roots and fruits, and much else. When a ploughman like Andrew Harvey was negotiating a fee at a hiring fair, into the bargain would go any and all of his children who could lend hands when needed.

Such crowded living conditions limited privacy, and it will have been difficult for Andrew and Catherine to be alone and undisturbed so that they could make love, especially in the winter. The unrecorded reality is that most couples, married or not, had sex out of doors. Lovers' Lanes and phrases like 'begotten in brake and bush' and the building of bowers are the memory of that social necessity. The wonder is that the Harveys managed to conceive six children, given the bad weather of the 1890s.

For the first time, there is a traceable continuity at the Henhouse. The Valuation Roll of 1885 shows that the Harveys were in the cottage at that time, and sixteen years later they were still living there. The 1901 census lists only two children who had not left home: the youngest, James, was eleven and at eighteen, his brother, William, was apprenticed to a tailor in Selkirk. Catherine's sister, Jane, had come to live with the Harveys and Andrew had changed his job. Having given up working with the heavy horses at the ploughing and carting, he had a job in the garden at the Haining. Perhaps at fifty-six his strength was not what it had been after forty winters in the fields. But not quite yet a 'done' man, Andrew could plant kitchen crops, dig potatoes and water the greenhouse plants.

In both the 1891 and the 1901 censuses, gamekeepers lodged with the Harveys, probably sleeping in the downstairs room in the south gable. Four adults and six children crammed into a tiny kitchen around the warming range will at least have had each other's body heat to supplement the glow of the logs. And in summer life will have been lived outside, as much as the weather and work allowed. Where they all slept in the tiny cottage is a matter for conjecture, but it is likely that all eight of the Harveys slept in the two upstairs bedrooms under the eaves. These were small, with an east-facing dormer window in each to give morning light. The younger children probably slept with their parents and the older boys kept each other warm in a big bed through the long, cold winter nights. Modern single beds are the nineteenth-century invention of hoteliers. Until relatively recently almost all beds had room for more than one and children were used to sleeping together for warmth and company, no doubt bickering over space and who was kicking whom, or pulling the quilt away.

Even though the 1890s in particular and the Edwardian period before the outbreak of the First World War saw the apogee of high farming, the children of the ploughmen and the farm labourers were beginning to drift off the land, leaving forever the *auld life* that Bina loved so much. Nicholas Wilson from the 1861 census in the Henhouse, easy to trace because of her quirky Christian name, moved to work in the textile mills in Galashiels, and William Harvey worked in a tailor's shop in Selkirk. By 1911 my great-grandmother, Annie Moffat, had given up the hard manual work of a bondager and moved into Kelso and domestic service. Bina became a seamstress because, as my mother used to say, she had clever hands.

Now only two of us live in the house that once rang with the clamour and chatter of generations of children, and sometimes I think it feels empty. No doubt Catherine Harvey would have liked that, wanting a moment or two to herself, some peace from all the demands made on her.

5 December

Walking in the winter dark, even on familiar ground, can be hazardous. I often set out with Maidie before first light and sometimes trip or stumble, occasionally splashing through an unexpected puddle. But this morning I fell headlong over a stone in the track that was definitely not there yesterday. Skinning the balls of my hands and banging my knee, I fell hard and jarred myself. But that was the least of it. The injury to my dignity was grievous even if no one witnessed the pratfall except Maidie. She simply stood and looked at me, my face suddenly at her level, cocking her head from side to side.

The dark landscape is lit by winter headlights swinging across the flanks of the western hills and on the road down from Greenhill Heights. Like Nicholas Wilson and William Harvey, these are people who live in the countryside going to work in the towns. But they do not have to walk, and risk injury.

6 December

No one could have known it at the time, but the 1911 census was a snapshot, a bright and sudden light on a world that was about to change utterly. Despite the glittering ceremonial of the Dehli

Durbar in December that year, when George V and Queen Mary were proclaimed Emperor and Empress of India, war clouds were gathering over Europe and cracks were beginning to appear in the vast British Empire. As the King-Emperor and the Queen-Empress sat in state under a gorgeously decorated canopy, their ermine and velvet cloaks cascading down the dais, they received the homage of native princes and potentates. But one was less respectful, and the Gaekwad of Baroda, Maharaja Sayajirao III, delivered a calculated insult. He approached their imperial majesties without wearing his jewellery and in a simple costume before bowing from the neck and turning his back to walk away from them. It was an omen, and thirty-five years later the Raj was dissolved and the British Empire began to crumble.

What accelerated this spectacular decline was the shock, wreckage and slaughter of the First World War. Britain lost a generation of young men, 'the best of them', and the nation's vitality and resilience seemed to drain away. Almost three hundred young men from Selkirk and the surrounding farms have their names carved on the war memorial in the town. Many died in the debacle of the Dardanelles, an abortive attempt to invade Turkey and attack Constantinople, the capital of Germany's ally, the Ottoman Empire.

In May 1915 my grandfather was home on leave in Kelso from the trenches in Flanders when he met my grannie Bina. She and Robert Charters had a brief affair that left her pregnant with my father. Soldiers knew that survival rates were low and an air of now or never may have swirled around them that summer as they walked down a lovers' lane. Who knows what promises were made. Soon after his return to the front, Robert found himself in a tunnel when the Germans filled it with mustard gas. Before the poisonous cloud reached him, he heard men screaming and somehow managed to crawl out. Badly blistered but not fatally affected, Robert spent many months convalescing, although he never made a complete recovery and remained a semi-invalid all his life. It was at that time he heard Bina was expecting his child.

More than a century later, and never having spoken to my father about his father, it is impossible to reconstruct the circumstances that led my grandfather to marry someone else, leaving my grandmother to raise her son with the help of Annie, her mother. War not only saw the slaughter of millions of men, it could also change the lives of women utterly.

7 December

Walter and his friend, Jean Wood, helped me to find Phil Cornwall. I was told that as a little boy he lived in the Henhouse at the end of the Second World War and now he has a house in Selkirk. Unknowingly, I pass it most days on my way into town. When I phoned Phil this morning to arrange to meet, he told me that many years ago he had walked down the Bottom Track to see the remade house. But when he met my daughters, Phil said it had been 'a wee bit awkward'. That is a shame. We shall talk on Monday and I shall invite him back to his old house.

8 December

In the night a rainstorm lashed the house and rattled the slates. By the time I was out and doing, the clouds had scudded past and across at the farm my neighbour had clanked his digger into the steading to lift steaming loads of silage into the cow byre troughs.

We drove over to Melrose to buy a Christmas tree, a journey that seemed much more recent than a year ago. Like birthdays, Christmases are annual signposts, accumulators of memory that loom up out of the mists of the future faster and faster. Up a winding road and down a farm track, trees were stacked, sheathed in white plastic netting. We asked for an eight footer that looked good bushed out. It turned out to be nine foot but we managed to fit it into the car, the top jammed between the front seats and brushing the windscreen. I changed gear by Braille.

When the tree was set up on its stand, Grace came over to help her grannie decorate it. Sitting nearby, watching, occasionally reaching for my hankie, I was much moved and taken back many decades to the time when our children were young and wide-eyed about Christmas. With great concentration, the three-year-old reached into old boxes, some dating from the 1970s, to unwrap glass spheres, ceramic redcoat soldiers, the star for the top, reindeer, glass bells and yards of tinsel and fairy lights. These decorations are as close to family heirlooms as anything else we might pass on. Some have been with us for forty years, others were bought on a visit to Berlin (where they understand Christmas and make very beautiful wooden carvings), and many are as old as our children.

The tree is a source of sparkling cheer, lighting the winter darkness, and it turns out to cheer others too. Across the Tile Field, in her cottage by the old Roman road, one of our elderly neighbours lives alone and last year she told us she smiles when she sees our Christmas lights twinkling in the darkness.

9 December

'I was born in your house.' Phil sat by a window with wide views over the Ettrick Valley, his own house perched high above the A7 as it winds downhill into Selkirk. 'My family moved in in 1941 and we left in 1951.' Just as with Bina, the place where Phil lived the first six years of his life stamped his memory indelibly. 'The stairs were awful steep. Have you still got them?' We do, and that remark came racing across the decades, the observation of a wee boy who probably climbed them on his hands and knees. 'Do you remember they were red-leaded?' I asked, and Phil nodded, turning away to look across the valley beyond his window.

His early memories of the Henhouse were sharp: a small cottage with no electricity and no running water. Dark winter evenings were lit by tilly lamps, 'very bright', paraffin lamps and candles, when they could be afforded. Instead of overhead bulbs clicked on by a switch, it was a childhood remembered in intimate pools of light and shadows behind, in the circle of firelight and in black darkness beyond the windows.

Water was drawn into the warm kitchen by a hand pump over the sink. About fifty yards to the west of the house was a well, the one plotted on the Ordnance Survey published after 1856. All traces of it have disappeared. Perhaps the springs upstream have been diverted with all of the building work and earth-moving done to rebuild the house and the new annexe. Water was heated in kettles suspended on the swee over the fire in the black cast-iron range and poured into a tin bath. No doubt the water was shared by several bathers and topped up to keep it warm.

Phil remembered that his parents slept in the kitchen and that the downstairs room on the left was a bedroom. Upstairs was another bedroom and a store. One of the four children avoided going into it because he felt 'a presence in the room'. All of these descriptions and recollections speak of a long continuity. Anne Moscript, the Wilsons, the Harveys and the Cornwalls all lived in a house that had changed very little since it was built in 1821.

And other links are remade. After Phil's birth in 1945 and the birth of Grace Moffat in 2016, a gap of almost sixty years has closed, as the Henhouse once again comes alive and nurtures children.

Vivid incidents lodged in Phil's memory. His parents used the old looseboxes adjoining the south gable of the cottage as a place to store wood ('I can't remember coal burning. How would they have got it to the house?') and also to keep chickens. The game-keeper at the Haining, 'not a nice man', had a polecat and it got into the loosebox and killed all the chickens before Phil's dad could catch it. When he killed the polecat, it bit through his wellies and the wee boy remembered the blood. 'When he brought the polecat into the house to show us, it stank the place out for weeks.'

In the long and drifty winter of 1947 the Henhouse was snow-bound for weeks, and food and other necessities were brought over the frozen fields by a horse and a sleigh. In all sorts of weather, Phil walked through the grounds of the Haining to school in Selkirk and walked back each afternoon. There was no Top or Bottom Track at that time; instead the cottage was reached by a path across what is now the Home Paddock. It sometimes shows up on early maps. There was no road or track for any sort of vehicle, except a sleigh. Phil's mum used to ask him to look for the Co-op travelling shop stopping up at Brownmoor so that she could walk up the path to Windy Gates, which was as close as the van could reach. Milk was left in a small churn down at Burn Cottage and one of the wee boy's chores was to walk through the fields and carry it back home.

In 1939 the Haining estate was put up for sale. As well as the mansion house, the farms at Hartwoodburn, Howden, Greenhill and what became Brownmoor were all up for auction. Phil's wife, Janis, had tracked down the particulars and a series of black-and-white photographs of the farms for sale. The Deer Park is parcelled up with the East Meadow, confirming its ancient, medieval extent, including the course of the Nameless Burn, a necessary water supply for the overwintering does.

Lot 30 is 'An Attractive Stone, Lime and Slate Detached Cottage known as Park Cottage'. On the accompanying map, I read clear evidence that we have been living at the wrong address. As Phil confirmed, the posh name and the postal address was Park Cottage. And yet the Ordnance Survey of 1900 clearly plots it as the Henhouse, as do earlier and later maps. Was the use of Park Cottage an attempt by the estate owners to airbrush a disreputable episode in the Pringle family's history? It certainly looks that way.

After the sale of 1939 and the great disruptions of the Second World War, the ownership of the Haining estate passed through three pairs of private hands. One of these attempted to evict Phil's family, four children and their parents, from Park Cottage. Clearly as tenants they had rights, perhaps even a long-term lease, but strong-arm tactics were tried. The little house stood in the open grass park with no fencing around it or the path to the Top Track, so the estate manager put a bull and some cows into the field. Since the only access the Cornwalls had to their house was the path, this was a clear case of intimidation. To keep warm and out of the wind, the cattle came right up to the front door and the walls of the cottage. Phil's mother went to see a solicitor in Galashiels and a fence was at last put up, but soon afterwards the family moved out.

It seems that the bullying new owner of the Haining wanted all of the cottages to be occupied by estate workers (Phil's dad worked for the county council as a lorry driver) but it appears that after 1951 Park Cottage was never reoccupied. Eventually the slate roof was breached and the pretty semi-circular porch vandalised when a previous farmer at Hartwoodburn cut away its supporting pillars to be reused as fence posts.

Phil never forgot the cottage where he was born and Janis gave me a copy of a photograph taken in 1989 with his mother and two children. Phil smiles broadly and they stand in the long grass with their collie panting on a summer afternoon. Two years later we drove down the track with another set of particulars, and immediately fell in love with this remarkable little house and all the spirits that seemed to swirl around it.

10 December

Yesterday afternoon I came upon a little bird sitting on the track outside the porch. It did not move as I approached and, thinking it was a chick (in December?), I picked it up gently, moments before the cat ran around the corner. With olive green feathers, short wings and a needle-like beak, it had remarkable markings on its head. Minute orange, black and yellow stripes ran from between its eyes to the crown of its head. In my cupped hands, its heart was trembling. To keep it safe from the cat, and in the warming, reviving sun, I put it on the roof of Lindsay's car.

While its downy feathers and tiny size had fooled me into thinking this was a chick for a few bewildered moments, I soon realised that the bird had probably clonked itself on the glass of the porch and was dazed. Nearly fatally. Looking through the *Book of British Birds*, I discovered it was a Firecrest, a close relative of the little Goldcrest that taps on the window frames of my office to flush out tiny, tasty insects.

11 December

Yesterday's rainstorms persuaded me to light both woodburners. High winds always find their way into old houses. But the log pile is dwindling rapidly and it will not last beyond the middle of January at this rate. Having heaved one of the thick lengths of Scots pine from leaning against the gable of the barn into the wood yard, I propped it over another thick log and fired up my chainsaw.

The pine was soaking, only the core dry. This presents me with a real problem because that wood had seasoned longest, having been cut in 2017 – it was my reserve supply. I cut the length into seven shorter logs and stacked six of them under cover to see how long they would take to drain. Meanwhile I put the axe through the seventh to see how long the split logs take to burn in a hot fire. They almost put it out and had to be removed.

12 December

Last night Adam, Kim and Grace came over to discuss plans for Christmas and we sat around the blazing woodburner as Lindsay noted down menus, likes, dislikes and a timetable. As ever, manufactured deadlines and crises will flare up between now and the arrival of Santa and, as ever, they will melt away. Quite why this Christmas tradition exists, I have never understood.

13 December

When the Cornwall family lived in the Henhouse, revolutions were gathering pace in the fields around them, the new obliterating the old with dizzying speed. At the beginning of the Second World War, the Haining estate and house were requisitioned by the army, first by the Welsh Fusiliers and later by the Polish Free Army, accompanied by their famous bear, Wojtek. The soldiers left two legacies. Under the soil of the Deer Park, buried in the surrounding fields and elsewhere, Rory Low has found the hidden debris of war, mostly bullet and shell casings, and on my desk sits the brass base plate of the Mills bomb, the hand grenade that he dug up not far from the ruins of the doocot. The second legacy was accidental.

One of the photographs in the sale particulars of 1939 shows the formal, neo-classical frontage of the Haining, with its row of six statues on their plinths between it and the loch. In the foreground, two men are boating near a circular, artificial island with a small tree growing on it. The Canaletto-style composition is ruined by a somewhat ragged-looking old building standing hard up against the west wall of the self-consciously stately mansion. Rubble-built with small windows and not a column to be seen anywhere, it looks as though a stray image from another age has somehow wandered into the photograph. And that is exactly what it is – the old house built by the Pringle family and expanded in the seventeenth century. All out of keeping with its surroundings, the crow-stepped gable end of the three-storey main building faces the loch, its frontage deliberately turned away from the view. Like the Henhouse and all of the farmhouses in our little valley, its entrance faced east, out of the prevailing west wind, keeping the winter draughts manageable. Shelter mattered more than a vista. Abutting the old house is a two-storey annexe and, in the angle created, what looks like a later porch. Apparently, it was used as servants' quarters. But in 1944, while the Polish army was in residence, the old house was burned to the ground, miraculously leaving the new undamaged.

In the fields beyond the loch and the policies, farming was changing. During the Second World War, it was essential that Britain was as self-sufficient as possible and farming was forced to adapt, largely successfully – although no one saw a banana for five years, rationing meant that no child or adult went hungry. In fact, general health improved. After the war, the policy of self-sufficiency continued but instead of increasing agricultural employment, workers began to leave the land. The *auld life* was being quickly dismantled as a direct consequence of the policies of the new and radical Labour government. Mechanisation was seen as the route to greater productivity and the 1947 Agricultural Act made that possible. By guaranteeing minimum prices for grain, meat, milk, eggs and all sorts of output, the government enabled farmers to invest in machinery.

The principle agents of change were the little Massey Ferguson tractors. Designed by an Ulster Scot, Harry Ferguson, they had a three-point linkage at the back end. This device enabled the tractor's engine to power whatever implement was attached behind. In the past, early tractors had merely pulled implements with more power but less precision than the big

Clydesdale horses. Ferguson's tractor became affordable and widely available after the war and well over half a million 'wee grey Fergies' were turned out by the Standard Motor Company in Coventry.

The effect of the Fergie was to bring the ancient skills of horseworking to an abrupt end and by the late 1950s it was unusual to see Clydesdales in the fields of Border farms.

As tractors pulled ploughs and harrows, farmers began to change radically the look of the countryside. More than half of all the hedges in the Borders were grubbed up and flower-rich hay meadows almost completely disappeared. The landscape lost much of its variety and texture, and its flora and fauna fled or died out. What the census called 'agricultural labourers' left the land in large numbers. More than three-quarters of the country population migrated to towns and cities.

Phil Cornwall told me that after his family left in 1951 no one lived in the Henhouse. When he and Janis came down the track on sunny afternoons to visit, it grew ever more ruinous after each passing winter, becoming the resort of pigeons and pheasants. Janis remembers finding many nests and pheasant eggs in the old kitchen and the downstairs bedroom.

14 December

The dawning light in the winter sky has moved around almost to south-south-east, and by the solstice in a week's time the sun will rise close to due south. I dislike the early dark and the closing down of the days. We have to bring the horses into their stables at 3 p.m., change their rugs, feed and settle them for a long night. After all these chores are complete, it is dark and I find it difficult to go back to my office to restart work. It feels less like laziness and more like a midwinter slowdown.

15 December

There has been no rain for four days and the bitterly cold west wind has dropped. With Maidie, I went out to the East Meadow to check on the Old Boys, the mares and the mini-Shetlands, knowing that we would not be wading through the clatch, the cloying mud of the last few weeks. I also wanted to look at the fencing and report back to Lindsay any breaches or bits missing.

Some years ago, we divided the seventeen acres of the East Meadow into five large paddocks. The farthest east is by far the largest and its undulations are good for the Old Boys. Not only do they stretch their arthritic legs up a few gentle inclines, one of the deeper dips offers good shelter from all but a north wind. And we also made a kink in the fencing to incorporate about ten yards of the course of the Nameless Burn. Fast-flowing fresh water rarely freezes and in all our time here that has happened only once. In the terrible Arctic winter of 2009–10, the temperature dropped to minus twenty for a few days and even though I managed to hack through the ice in the burn with an old axe, it froze again each night. We had to take out big water containers and pour the contents into buckets for the horses to drink there and then.

In the middle paddocks, we had a large field shelter built for the mares. In periods of bad weather, especially wind-driven rain, they also seek refuge in the lee of the sitka plantation at the top of the meadow. In their paddock on the flank of the Deer Park, the minis have no bield against the winter winds. Their two-ply coats keep them warm.

Perhaps what attracts me to this place and its fields is an ancient sense of harnessing the land, of growing food for people and animals, and, where we can, encouraging the return of the flora and fauna that fled in the face of the revolutionary changes in farming after the Second World War. We have planted hundreds of trees, left acres of wetland around the Nameless Burn untouched and run about a mile of hedging around the home paddock and elsewhere. So that birds and small mammals like hedgehogs, stoats and others come back, we mixed hawthorn, some holly, hornbeam, crab apple and blackthorn. The latter has turned out to be very invasive, sending shoots and suckers underground into the paddocks; its vicious long spikes are dangerous for horses, but it is also good protection for small birds from their predators. In full summer lustre, leafy and dense, the hedges are very beautiful and brim with life. When October winds blew away the last of the leaves, a perfect blackbird's nest was revealed, its twigs interwoven in a delicate, symmetrical tracery, something that had been completely hidden all summer as eggs were laid and chicks hatched and fledged. Even the frosted winter fields looked welcoming this morning.

16 December

At eight hundred feet, another world glowed pale white in the half-light of dawn. On the uplands behind Hartwoodmyres and the Thief Road lay a covering of snow, and yet here, two hundred feet lower, there was nothing except a sheeting of thin ice. The world looked like a Christmas cake topped with white icing. I hope we stay down amongst the sultanas, the mixed peel and the almonds.

The two living children of this place met this morning. When Adam brought Grace to walk the big dogs, Phil Cornwall arrived at the same time and they greeted. The three-year-old wee lass did not turn to her dad in shyness and they smiled at each other. I felt the tears prickle.

Janis Cornwall told me that Phil had always wanted to see how the house where he had been born had been brought back to life and so we began by walking around the outside. Not much seemed familiar and so many trees, shrubs and hedges had been planted that the formative geography of Phil's first years had been so radically reformed that he did not recognise much of it, only the vistas up the little valley.

The ground floor of the old cottage and the large extension to the west are all open plan and Phil found it a little disorienting. 'But I like what you've done. In fact I am a wee bit envious.' When I suggested he sit in the corner of what had been the kitchen where the family ate and where his mum and dad slept, I sensed that blurred memories were at last beginning to loom out of the past. To his left he could see the familiar view to the Tile Field and Hartwoodburn Farm, but mostly he seemed to recognise that he was sitting by the same window he had looked out of as a wee boy and by two walls that had bounded his world. The steep stairs had stuck in his memory and the step down to the ground-floor bedroom where his sister slept. It was as though an old and faded black-and-white photograph had been torn up and only a few shreds and edges of the picture remained.

Janis sent me a photograph she took of Phil and me standing outside the house. Both of us were smiling, knowing that a circle had been completed. Phil seemed very happy that the old ruin had come back to life and accepting that its revival had submerged much of his past. But it was always so in the long history of here. I gave Phil and Janis a prehistoric arrowhead and a spindle whorl that the Mason brothers had picked up from the fields around the house and that seemed to cement another continuity. Just as the lost lives of the hunter-gatherers had laid down an early layer of history, so Phil's early years in this place had been laid over many others in between. And his experience here has now been overlaid.

18 December

Back to Edinburgh for some legal business involved in winding up a charitable trust, but this time I used the train journey to do some good work. Sometimes, the structure of a book can take a long time to fall into place, and a new project seemed to click together like the pieces of a jigsaw today. Christmas is coming and I have only one or two meetings before life begins to revolve around shopping, cooking, eating and drinking. And we all forget which day it is.

19 December

To the Northumberland coast and the fishing village of Craster for Christmas lunch with my sisters and their families. Not only is it a handy halfway point (they live in Newcastle) to meet and exchange batches of presents, it is also a gust of salt air off the chill North Sea, something very welcome in my landlocked existence. And the sea was wild this morning, white breakers crashing against the harbour wall. Before lunch, I shopped at the famous kipper smokery for salmon, cod and, of course, kippers, all strong tastes of the sea in the deeps of the winter.

I enjoy the journey down the Tweed Valley and then the A1 to the south. Between it, the main railway line to London and the coast is a richly detailed landscape of farms, grand houses, dense native woodland, worked-out quarries and coal pits, very different in atmosphere from Scotland and yet so close to the border. I drove back through Seahouses, a miniature Blackpool of neon-lit amusement arcades, chip shops and gift emporia, before reaching Bamburgh and its majestic castle. Restored by the industrialist and arms manufacturer William Armstrong, it was originally the capital fortress of the ancient kingdom of Northumbria, a seamark rock visible along the coast and far out to sea. Then I drove past the wave-washed Farne Islands and magical, other-worldly Lindisfarne before turning inland for home.

20 December

This was what my mother called a back-endy day, one of those dreich, dowie, listless days at the back end of the year when nothing seems to get done. Recovering from a bad cold and with a thick head, I could not seem to get on with anything, and disastrously before Christmas I woke up to find I could not taste anything. And so I chopped some logs and did some household chores. Potentially most satisfying was to pour kettles of boiling water over wax-encrusted candle-holders to clean them. But the sink ran red as I cut myself badly when one of the glass sconces shattered.

21 December

Today is the winter solstice! This is the turning day of the year. The sun will rise, most probably hidden behind horizon clouds, at 8.38 a.m. and it will set at 3.41 p.m., allowing only seven hours of daylight and plunging us into seventeen hours of darkness. But tomorrow, edging across the southern ridge of Greenhill Heights, the light will begin to return, and after two weeks of the festive season it will seem that we have put some distance between ourselves and the deep midwinter dark.

This morning I saw a long sliver of pale yellow light along the south-eastern horizon and it cheered me. The sun is hope: as the morning breaks, we wake and work begins. I think of that golden light in the east as Lindisfarne light, glowing from the little island on the edge of Heaven. Walking back with Maidie, I saw smoke pluming above the house from the fire I had lit an hour before. Damp logs burning slowly. The windows twinkled with warmth as I carried in another basket of drier firewood to get a blaze going.

22 December

Last night candles lit the past and made the future glow. Around a table set for a celebration sat three generations of our family. The faces of our granddaughter, Grace, and all of our three children and their partners were warmed by the gentle flicker as dishes of ham, roast potatoes, buttered leeks, peas, boats of gravy and, later, jugs of cream for sticky pudding were passed around amid a hubbub of happy chatter. I can't remember what anyone said. I didn't need to. All the children of this place had gathered together and what will stay with me is the warmth of that evening.

Candlelight is like the ancient circle of firelight, with darkness behind, a focus for stories, for the making and remembering of family lore. Eight thousand years ago, no longer a time out of mind for us here, fathers and mothers sat around the crackle of flames and celebrated the solstice, the longest night. Our ancestors' calendars may not have had dates and months, but they knew the year's turning times, when the days were shortest and longest. There is some slight evidence that they counted the passing of time in nights rather than days. That habit is recalled in the peculiarly British word 'fortnight' for fourteen days and the now obscure 'sennight' for a week.

The mothers and fathers who sat in the circle of firelight in their shelter by the shore of the Hartwood Loch were very rarely joined by their grandchildren. Most died too young to see their children have children. When farming came to our valley about five thousand years ago, lives became more settled and the population expanded very quickly. In their snug roundhouses, as the smoke spiralled into the conical thatched or turf roofs, some men became grandfathers and those few women who survived the perils of childbirth and the damp of too many winters will have seen a third generation begin to unfurl after them. That new sense of continuity linked to the ownership of land changed how families saw themselves. The passing on of customary rights and agreements became more important. And as hierarchies developed, families became part of wider kindreds. Our valley gradually became home to a community.

Candlelight seems to collapse time. The habits and fabric of the lives of those who lived on the land remained largely unchanged for many millennia, but the last three generations of the four hundred who have passed their lives in this place have seen seismic shifts in how society is organised, from the universal availability of electricity and modern medicine to the digital revolution. The flicker of candlelight and the flames of the wood-burner transport us back beyond those shifts to a time of stories and simple bonds. All that experience in this place is not lost.

Last night our family told itself familiar stories of a shared past and obligations freely given, of complexity and unexpected turns. But mostly we told stories of love, sometimes awkward, sometimes unarticulated, sometimes surprising, but stories of love nonetheless. If the memory of these precious times is all that survives of me, I shall die content.

23 December

We have a week or two of hard work ahead of us, as we muck out our stables and look after all sixteen horses without help. We fed the Old Boys and the mares yesterday, and I enjoyed being out in the East Meadow doling out the hard feed and haylage. Gem, the oldest at thirty-two, stayed at the far end of the meadow despite calls and me rattling the buckets. He is probably deaf now, but by tomorrow or the day after he will see the others move and follow them.

24 December

Last night ghosts stalked the Long Track. Suddenly the armoured knight rode out of the darkness of the past, his destrier's long caparison billowing, its blood-red crusader cross bright against the white silk.

In a grassy hollow on the old west road that runs to the Haining behind the woods bordering the Doocot Field, Rory has found a sword pommel. It came from the hilt of a heavy longsword of the sort wielded by knights and dates to the late thirteenth or early fourteenth century. It is almost identical in shape and style to another pommel found recently at Ashkirk, three miles to the south, on the old road to Hawick, the continuation of the Long Track. What lit up this find and links it to Rory's are two inscriptions carved on the polygonal surfaces. *Zion* was another name for Jerusalem and *INRI*. It was daubed on the titulus, or name plate, hammered onto the top of Christ's cross. The acronym translates as 'Jesus of Nazareth, King of the Jews'. Along with a series of plant motifs, probably hawthorns to symbolise the crown of thorns, the inscriptions mark these pommels as relics of a savage piety and were believed to impart magical, protective powers.

Proclaimed in 1095 by Pope Urban II, the First Crusade set out from Western Europe to recapture Zion, to take back the sacred places of Jerusalem and the Holy Land from the infidel. In one of the most remarkable military and political episodes in all history, thousands of Christian soldiers marched and rode east to defeat Muslim armies and establish the crusader kingdoms in what is now Palestine, Israel, Lebanon and Syria. In fierce heat, squadrons of armoured knights charged with their destriers into the hosts of the heathens and scattered them. Their courage and audacity were sharpened by the certain knowledge that, as crusader-pilgrims, their sins were all shriven and, should they die in the Holy Land, their passage to Heaven was assured.

As these outposts of Christianity in the east inevitably began to shrivel, more crusades were preached by the Popes in Rome, and in 1270 Lord Edward, the eldest son of King Henry III of England, sailed to the Holy Land and fought back the tide of Islam for two years. Those knights who rode with the future Edward I wore the cross as a badge of piety and renown ever after. And it seems that some of these warriors for Christ rode up the Long Track in July 1301, when the king invaded Scotland and came to Selkirk Castle. They had followed him in the searing heat of the deserts of the east and now they came north in his service.

Sources hint at the identity of those who sailed with Lord Edward from southern France in 1270. Drawn from the wide

European connections of the Plantagenet kings, it was a cosmopolitan army whose lingua franca was almost certainly French. And despite the fact that his father had given Edward an old English royal name, Edouard looked to Europe as much as to England. His mother was Eleanor of Provence, daughter of Beatrice of Savoy, the western alpine province of what is now Italy. It was from the fertile plain around Turin that Edward gathered a group of knights who may have been seen as an inner circle, perhaps even royal bodyguards.

Chief amongst this group of adventurers and closest to Edward was Otto de Grandson, sometimes written as Grandison. It was said that no one could do the king's will better, including the king. When an assassin sent by his enemies in the Holy Land stabbed his master with a poisoned dagger, Otto sucked out the poison and spat it on the ground, risking his own life. Others in the group known as the Savoyards were Otto's son William and Sir William de Cicon and his brother, Sir Stephen.

When the Ninth Crusade finally reached the Holy Land, they made little impression. By 1270 the coastal cities were all that remained in the hands of the squabbling crusader dynasties and Edward had only a thousand men, including two hundred and twenty-five armoured knights. When Henry III died in 1272, his heir returned to England to claim his throne with his Savoyard knights, all of whom had earned the right to wear the cross.

In the 1280s Edward began the conquest of Wales. Otto de Grandson was appointed Chief Justiciar to rule on his behalf and he was based at Caernarfon Castle, the key fortress in the north. This remarkable structure, modelled on the Land Walls that defended Constantinople, was built under the direction of the master-mason Walter de Hereford. Records show that in 1300 de Hereford and his workforce were in Carlisle, on their way north with Edward to build fortifications as his army conquered Scotland. This was the same strategy of castle-building consolidation used to subdue Wales.

It is very likely that Otto de Grandson led the contingent from Caernarfon and that their expertise turned Selkirk Castle into a formidable fortification. It is a leap to assert this, but well within the bounds of likelihood that some of the Savoyard crusaders were encamped in our fields in July 1301, and that in an exhibition of swordplay in a grassy hollow by the old west road one of them either lost or dropped the pommel of a long and heavy sword, or it was damaged and discarded. Seven centuries later Rory Low picked up this far-travelled relic of the Ninth Crusade.

I have written this diary in real time, only going back through its entries to cut repetition and tidy up as best I could. When I wrote a year ago in the introduction about an armoured knight, riding a destrier, wearing a caparison with a crusader cross blazoned on it, all I knew was that the Long Track was old, almost certainly medieval. The knight was only an image, an intuition, a symbol. I had no idea that at the end of the year Rory's skill with a metal detector would make the knight real, a warrior who had felt the fierce heat of the deserts of the east on his back and who rode up the Long Track to have his pavilion set up in our green fields.

With characteristic insight, Walter asked if the hollow where the pommel was found had an enclosure around it and was a good place for soldiers to watch. 'Yes, it's like a little theatre,' said Rory. Walter's hypothesis was that the pommel was discarded or lost because a sword may have been damaged during swordplay. Knights were professional soldiers and often practised their skills. Perhaps an older crusader was taking on a younger man. Perhaps silver pennies like those Rory has found nearby were wagered on the outcome. The pommel may be a long echo of the clash of steel on the old road. It was watched by a ring of spectators by the waters of the Ettrick as they remembered Zion and the knights who defended the gateway to Heaven from the infidel.

25 December

A happy day to remember events that took place in Bethlehem, not far from Jerusalem, two thousand years ago, the gentle, touching story that fired the fierce piety of the crusaders and inspired peace and love as well as war.

Christmas carols are amongst the most resonant of hymns and I cannot get 'Hark the Herald Angels Sing' out of my head as I work in a stable that is not so different. Three kings will probably not appear, but my children and grandchild will.

This morning a chevron of about twenty geese honked loudly as they flew over the farmhouse, much better than French hens or turtle doves.

26 December

The ancient rhythms of the past have returned. The hard work with horses begins at first light and ends in darkness, as they are fed in their stables in the evening. When at last we sit down in front of the blazing woodburner, a comfortable seat is seductive. It would be very easy to stretch out my tired legs and fall into a doze. But if we did that, the days would begin to become chaotic, have no structure. The dogs and the kitten, as well as the horses, need looking after and so up we get with more to do. In the farm cottage at Cliftonhill, in the harsh winters of the 1890s, the same grunts and groans would have come from William Moffat as he stood up, put on his boots and walked over to the stables to give his Clydesdales their evening feed.

27 December

Perhaps because I do it every day on the farm, I have taken to walking as a vehicle for telling stories. Walking through history, the places where it happened – and it happened almost everywhere – seemed to make it come alive or at least draw the past closer. My last three books have all been journeys of one sort or another, and while I try to arrive at some conclusions, they are usually provisional, somehow part of a longer journey, and not always associated with facts, events and dates.

Walking and thinking, sometimes dreaming, sometimes flying far above the landscape, the clouds scudding by, I often drift, not focusing on anything for more than moments. I used to see such flights of fancy as a waste of time, but now I am not so sure. At the end of a walk, after the metronomic, hypnotic rhythm of putting one foot in front of another, I feel complications begin to untangle. My fancies sometimes turn out to be something more; intuitions, occasionally ones that solidify into facts, particularly when I walk exactly where saints, kings or artists walked and I see what they saw. And it sometimes offers a different gloss on what they wrote or was written about them.

In the past, I used to spend long weeks and months at my desk reading. I began with primary sources where they existed, moving on to other history books about what interested me, trawling the internet for good material, and when I felt I had exhausted all of those, only then would I sit down to write. But now I pull on my boots, put a notebook in my pocket, sometimes a map, always a phone camera, and I go looking for ghosts, listening for the jingle of harness, the creak of cart wheels, the songs of field workers, the peal of ancient bells. If you look and listen quietly, the land will whisper its secret history.

I have also taken to writing it down as I find it, as I go along, and when it seems that a journey has ended (it is rare for me to think it is completed) I try to organise a narrative. The most attractive aspect of looking to make sense of the past in this way is that I am often surprised; I often come across a notion that would never have occurred to me as I sat at my desk.

This diary is a record of things I came across in a tiny area, no more than eighty acres of hill, field and stream in the Scottish

Border country, and yet it has turned out to be a much richer and wider story than I could ever have anticipated. The whispering land has given up tales of hunter-gatherers and the prehistoric fauna of the painted caves of the Pyrenees, of Irish lords who marked their territory in stone, of crusader knights and the wars of the Holy Land, of the War of the Three Kingdoms, of the fall of the Stuart monarchy, the agricultural revolution, of romantic Polish soldiers, world wars and much else. The debris of history was strewn everywhere; these stories were sparked by a flint arrowhead in a furrow, a cigarette lighter lost in a field. None of this was predicted or premeditated, it just happened as I walked and took flight when Rory and Walter walked with me.

History should not be thought of as rows of books in a library or a bookshop, or the subject of drowsy double periods on a Friday afternoon. History is us, it is what made us, the only reliable means of understanding the present or making any sense of the future. And all that is needed to make it come alive is curiosity, and a pair of stout, waterproof boots.

28 December

Walter Elliot likes to dream of the past. A child of the Ettrick Valley whose eighty-five summers have been spent in or near the hills and valleys of the high country, he can read the land instinctively, understanding how it was seen by the hundreds of generations before us. He taught me the importance of vantage and visibility. When we went with divining rods to look for the Anglian settlement at the north end of the Doocot Field, he knew immediately where to begin. Without a word, Walter walked up the slight slope to a ridge I had not noticed. In the millennia before good drainage, and in a landscape with many fewer trees, people often preferred to build and live on higher ground, and on vantage points that offered good visibility of the surrounding countryside. With the divining rods, we came upon the outlines of old buildings in only a few minutes.

The patient quartering of likely ground, the precise recording of find-spots and an encyclopedic knowledge of old coins, medieval in particular, has meant that Rory Low's discoveries have complemented Walter's perceptions and conjectures perfectly. When the metal detector buzzes and an object is brought back into the light, history finds its way into the palm of Rory's hand. Coins, brooches, the base plates of hand grenades, a sword pommel, lumps of lead – all of these objects carry the fingerprints of the past. The last person to touch them before Rory was the person who lost them – in the seventh century, the thirteenth, the seventeenth and the twentieth. Edek and Rita, Otto de Grandson, Edward I and a little boy who could not find his toy cannon in the Doocot Field all lift the veil between us and their long-ago lives.

This account of a year in the valley is dedicated to Walter and Rory not out of good manners or even gratitude for sharing all they know, but because they made many of its pages sparkle and sing of the past.

29 December

Eight days after the solstice, time seems to turn slowly, like an old-fashioned record player running at 33 rpm for a 45 rpm single. The words slur into echoic slow motion, so slow that no forward movement seems possible. Perhaps after the year turns the days will stretch out and the tempo will brighten.

30 December

At the foot of the Long Track, where it meets the C road, winter has not been kind. Despite our neighbours at Burn Cottage putting down tarmac planings, many puddles have turned into wide, sharp-edged holes. And the junction of the old track with the smooth road surface has a broken seam of potholes, humps and frayed edges. Perhaps that is as it should be at the precise place where the past bumps into the present. Roman legions marched past that junction of history, armoured knights and their esquires reined their horses there, and David Leslie's covenanting cavalry passed silently by on their way around Howden Hill to charge into the rear of Montrose's royalists. Rough edges are no more than appropriate.

The foot of the Long Track is also an annual beacon of warmth and conviviality, a welcome light in winter. Usually in the after-noon of the day before Auld Year's Night, our neighbours at Burn Cottage hold an old-fashioned soiree. All of the people in our little valley are invited to this Christmas-lit house, where tables groan with food and the pantry shelves and fridges are full of bottles, some of them with wine in them. Held between 2 p.m. and 5 p.m. so that guests can come and go before the descent of the black-dark of midwinter, this party can be a review of the year. Conversations often revolve around change and plans. The year that is passing is reviewed and the one to come discussed. Tomatoes figured in one exchange, how the year's crop was and what needed to be improved, the best varieties, the dangers of over-watering, growing from seed as opposed to buying potted plants and much else.

I do not think of these everyday things as trivial, just everyday, the stuff of lives, the routines that sustain us, the work we do and our effort to be as productive as possible. What Walter Scott called 'the big bow-wow stuff' also gets a careful airing at the soiree in discussions of politics, and this year I noted more talk than ever about the climate emergency. On this morning's news sites, a headline jumped out at me: 'Thousands flee to the sea as fires race to the ocean'. As temperatures soar to new highs, millions of acres of woodland and bush are burning in Australia. Across New South Wales and Victoria a 310-mile-long blaze is being fanned by winds and is impossible to put out. The news reports carried pictures of red skies and ash-filled air, as frightened people wearing facemasks fled to the beaches and quaysides. I did not have to scroll down far to find the word 'apocalyptic'.

Perhaps attitudes in at least one country will be shifted, but equally it may be that more apocalypses are needed elsewhere. Strangely, I am becoming slowly more optimistic about combatting the climate emergency. Action to save our planet will not come from government or big business, but from below, from voters and consumers, from the terrified people on the beaches in Australia, and they will force change.

31 December

In the Deer Park, Adam and I came across a symbol of royalty. It scampered up a bank above the Nameless Burn and then disappeared into a hole in the old ash tree that had had its crown broken off by a storm in the summer of 2018. And then, appropriately, it poked its head out of the top of the hollow trunk where the crown should have been. Adam had spotted a pure white ermine, a stoat whose coat had turned from brown to its snowy winter colours. Very clearly visible against the grey wrack of the ruined tree and the dieback around the burn, the ermine seemed busy, preoccupied, perhaps even ferreting around after something. Adam reckoned that its territory probably included the southern flanks of the Deer Park and especially the track that runs along the fences of the East Meadow. He has seen many voles darting into the long grass, the stoat's principal prey.

Because ermine pelts are small, the white fur was mostly used as a trim or a collar for ceremonial robes. Not only monarchs wear ermine, it is favoured by Popes. Many portraits show capes known as *mozzette* and berets fringed by the stoat's winter fur. White is, of course, intended as camouflage in the snow, and perhaps the busy little ermine knows more than the Met Office about the winter we can expect next year.

It is the last day of the old year, and the last entry in this diary of a year in our valley. A brilliant sun shone all day, and after darkness fell the light in the western sky glowed pale blue for a long time, stencilling the black line of the horizon I have come to know so well. Out in the early evening with the dogs, I began to think of the first time we saw this place. In the summer of 1990, we bumped down the Bottom Track and parked outside a grey ruin in the corner of a field. Biscuit-ripe barley billowed in the warm breeze, growing to within a few yards of the front door. The roof was caved in and the old porch half-collapsed. Inside was chaos covered in pigeon shit and pheasant droppings. The advertised stables had all but fallen down and the garden was a jungly mess.

We had looked at many properties in the Borders, some on the edge of villages, others very remote, hidden in hill valleys, and one was even in England, just over the border. But this unloved, abandoned ruin at the bottom of a bumpy track whispered to us. My wife and I exchanged looks and began walking around the old house, barely able to see where we were going in the overgrowth. Neither of us were thinking much about what it might become, about what estate agents call 'potential'. Instead, without much prior knowledge, we began to intuit what this place had been, began to be aware of its spirits. Most of all we immediately understood why a house had been built in this out-of-the-way corner. It was not only the wide and long views, especially open to the south, it was also its place at the foot of a slope, above the banks of a little stream. The house felt like a destination, somewhere we sensed we should like to stay.

The Henhouse took a long time to rebuild and was finally completed in May 1994, but in the intervening summers we often came down from Edinburgh, sometimes with a picnic on a sunny Sunday. And, slowly, we found that we did not want to leave, return to the noise and the bustle of the city. Even before a brick was laid or a nail hammered in, we knew that this was our place, somewhere we would make together, somewhere we would live, perhaps permanently. In those summer afternoons, the peace of the old house and the fields around it began to settle on us.

When building work began in 1992, we came down to see what progress had been made (too often none) but always stayed longer than we meant to. One afternoon, I thrashed through the tall nettles and willowherb at the bottom of the garden and, close to the banks of the burn, I slipped on a stone. After clearing away the tangle, I saw that it was a big, oblong stone that had once stood upright. In falling, it had kicked out earth along one side of a small slot where it had originally been set. With a fence post and a shovel, we heaved it up and it resumed its position, slipping easily into the slot.

The stone started me thinking. Was it a small, prehistoric standing stone? Was it a boundary marker? It was the first time I dreamed of what this place had been, the first time I heard the rustle of the leaves, and the first time I looked over my shoulder.

Acknowledgements

Simon Thorogood was a vital guiding hand in the writing and shaping of this book and I want to thank him very much for all his creativity and tact. His skill as an editor made a big difference.

Walter Elliot and Rory Low added great sparkle and their willingness to share their knowledge and perceptive judgements make them as much the authors of this book as I am. But I should say immediately that any mistakes are mine alone.

Andrew Crummy came down for a day on the farm and the beautiful drawings and map he made adorn the book.

My agent, David Godwin, kept faith with this project as it developed and changed. Thank you, David.

I always carry a notebook and, being forgetful, write things down when I see them or think of them. On several pages there is a long line as my pen slid off and often fell to the ground. These scores on the paper are a record of the times when my dog, Maidie, saw a rabbit and yanked the lead hard, sometimes almost pulling me over. But I would like to thank my little Westie for all the times she stood still while I scribbled something about the weather or the land.